AGRICULTURAL ACAROLOGY

THE AUTHORS

Dr. B.S. Chhillar is Director of Research at CCSHAU, Hisar. He received his B.Sc. (Ag. & A.H.) degree from PAU, Ludhiana in 1968, his M.Sc (Entomology Hons.) degree from HAU, Hisar in 1970 and his Ph.D (Entomology) from HAU, Hisar in 1980. At the CCSHAU, he also served as Associate Dean Post Graduate Studies, Professor and Head, Department of Entomology. In addition, he served as Secretary, Board of Studies of College of Agriculture. He has a vast experience of 35 years in the field of entomology in research, teaching and extension. He has been a pioneer worker in sugarcane entomology. He has about 50 research publications, 12 books/ compendiums, 5 practical manuals and other extension articles to his credit and more than 20 PG students. As Director, Centre for Advanced Studies (ICAR), Department of Entomolgy, CCSHAU, Hisar, he has organized eight national level trainings of three to four weeks duration each for scientists of SAU's and ICAR institutes.

Dr. Rachna Gulati is scientist working in the Department of Entomology, CCSHAU, Hisar. She obtained her Ph.D on storage mites and guided PG students on mites. She has worked with mites since 1990. She has experience on various aspects of acarines particularly storage, plant, honeybee and nematophagous mites. She has over 40 research publications, one book as co-author on grain storage (CSIR, New Delhi) and one compendium on entomological techniques and several extension articles.

Dr. Praduman Bhatnagar is Senior District Extension Specialist cum Co-ordinator, KVK, Hisar. He also served as Entomologist (Pulses) in the Department of Plant Breeding, CCSHAU, Hisar and Senior Scientist (Entomology) at Central Institute of Agricultural Engineering (CIAE), Bhopal for a brief period. He has been engaged in research activities in different crops for the last 12 years. He has experience of working in veterinary entomology and extension entomology. He has 28 research papers, 5 chapters in a monograph and 110 extension articles to his credit.

AGRICULTURAL ACAROLOGY

B.S. Chhillar
Rachna Gulati
Praduman Bhatnagar

2025

Daya Publishing House®

A Division of

Astral International Pvt. Ltd.

New Delhi – 110 002

First Published, 2013
Reprinted, 2025

ISBN 978-81-7035-477-2 (HB)

Publisher's Note:

Every possible effort has been made to ensure that the information contained in this book is accurate at the time of going to press, and the publisher and author cannot accept responsibility for any errors or omissions, however caused. No responsibility for loss or damage occasioned to any person acting, or refraining from action, as a result of the material in this publication can be accepted by the editor, the publisher or the author. The Publisher is not associated with any product or vendor mentioned in the book. The contents of this work are intended to further general scientific research, understanding and discussion only. Readers should consult with a specialist where appropriate.

Every effort has been made to trace the owners of copyright material used in this book, if any. The author and the publisher will be grateful for any omission brought to their notice for acknowledgment in the future editions of the book.

Published by : **Daya Publishing House®**
A Division of
Astral International Pvt. Ltd.
– ISO 9001:2015 Certified Company –
4736/23, Ansari Road, Darya Ganj
New Delhi-110 002
Ph. 011-43549197, 23278134
E-mail: info@astralint.com
Website: www.astralint.com

Digitally Printed at : Replika Press Pvt. Ltd.

Foreword

Agriculture has tremendous impact on human economies. But with the changing cropping system and emergence of mite as a serious pest, a 'let-alone' approach to its management will prove inadequate in meeting the demands for an abundant and high quality food supply. This concern stems in part from increasing incidence of phytophagous mites in various agricultural crops. Currently, 17 species of mites have attained the status of major pests and 30 as minor pests. Still others are on the verge of becoming pests in cash crops.

Increased awareness about the role of stored/ house dust mites as allergens and soil mites as bio-indicators has generated a renewed interest in their culture and management. The rapid spread of parasitic mites in mammals, birds and insects, especially honeybees, has led to decline in host population.

Acarines are of considerable economic significance and can threaten the sustainability of agro-ecosystems. There is an urgent need to create awareness about acarines and their behavior in different ecosystems so that timely management of harmful species and utilization of beneficial species can progress hand in hand. The authors have made industrious effort and collected exhaustive information on Acarology and presented it in a systemic manner. The book is divided into three sections–'Acarine biodiversity',

'Acarological techniques' and 'Acarines as biocontrol agents'. It is my fond hope and trust that this book will fill a long void in acarological research, teaching and extension. I compliment the authors for this significant contribution in the field of Acarology.

J.C. Katyal
Vice-Chancellor
CCS Haryana Agricultural University
Hisar

Preface

The science of acarology needs distinction from entomology and arachanology. There is an urgent need to make awareness about acari, which are gaining considerable economic significance. On one hand, phytophagous mites as pests of field crops and mites associated with granaries and warehouses cause heavy financial losses, other species have attained prominence as household pests and causative agents of allergic reactions in human beings and other mammals. Many parasitic forms act as vectors of disease causing organisms. With the appreciation of the probable importance of micro arthropods in maintenance of soil fertility, amongst which mites constitute a dominant fauna, ecologists are directing their attention to this group of animals. Acari are among the major resources for biological control of various classes of pests including insects, nematodes and weeds.

These widely distributed microscopic animals ranging from free living to parasitic forms and terrestrial to aquatic forms require different techniques for handling. In many instances, lack of information about the correct identity of mites and adequate knowledge regarding their biology and ecology, have hampered our ability to effectively combat these mite pests. The literature regarding their collection, rearing, biology etc. are scattered and most of the time unavailable to scientists interested to work on acarines, is a major drawback of acarology although acarological research and post graduate teaching exist in many agricultural and traditional universities as a branch of zoology/ entomology. As a sequel to these developments, a comprehensive book on agricultural acarology is prepared which covers all aspects of acarine biodiversity, acarological techniques and their role as bio-control agent in an abridged form. The book has been liberally furnished with illustrations for better understanding.

Part I is concerned with Acarine biodiversity which contains six chapters, one each on plant, stored, soil, house dust, parasitic and water mites. These chapters provide information on distribution, morphology, life cycle, nature of damage, pest status and management of mites in different habitats. We are heavily indebted to AINP on agricultural acarology publications related with phytophagous mites. Acarological techniques are the focus of Part II. Details on techniques of collection and extraction, preparation and mounting, labeling and housing the collection, culturing, counting and bioassay are given in individual chapters. There is recognition of acarines as effective bio-control agents of insects, nematodes and weeds because of their voracious nature and well developed searching, dispersal mechanism. Information related to predatory potential of acarines is included in Part III in three chapters.

An extensive list of references and glossary are provided at the end of the book as part of our effort to make information on these topics more widely available.

This book is primarily for students, agricultural scientists, extension specialists, crop growers and others with an interest in Acarology. Finally, in this book we hope to engender recognition for all Acari–terrestrial or water, phytophagous or post harvest associated mites, parasitic or predatory–and the potential of Acari as bio-control agents of pests. Due to scattered literature on techniques related with diverse habitat of Acari, scientists and students faced difficulty in working with mites. The purpose of this book is to provide a starting point to future acarologists and promote them as large, dependable workforces who do not have technical training in this regard

B.S. Chhillar, R. Gulati, P. Bhatnagar

Contents

Part II: Acarological Techniques

Part III: Acarine as Biocontrol Agents

Chapter 1

Introduction

Today's agriculture, which is an essential part of our economy is confronted with the shifts in pest pressure, rise in mite pest problems, stringent environment regulations and need for profitable high quality produce. Outbreaks of mite pests and more number of acarine species getting the status of pest is alarming as more than 40,000 acarine species are known from diverse habitats. Acari, because of its small size and cryptic appearance, are difficult to detect and hence most of the time infestations are overlooked. Once established in a new area, certain biological characteristics like high fecundity, various modes of reproduction, short life cycles, a myriad of dispersal techniques and adaptability to diverse ecological conditions allow rapid escalation to pest status. In the current scenario when world trade is showing exponential increase, these traits are putting mites on a stage of considerable economic significance where devastating situations may occur to threaten the sustainability of agro ecosystems. There is an urgent need to make awareness about acarines and their behaviour in different ecosystems so that timely management of harmful species and utilization of beneficial species can progress.

These small but mighty creatures attracted the attention of man when these were recorded as vectors of some of the most devastating and lethal disease agents; the classical example being the use of chigger mites in World War II which led to the death of thousands of soldiers.

1.1 Development as Plant Protection Science

The science of Acarology, which includes mites and ticks, needs distinction from Entomology and with in Arachnology. Acari are considered among the most successful arachnids of universal distribution which can be measured on the basis of greatest diversity shown from free living to parasitic forms, univoltine to multivoltine species, may live in the germ of grains, leaf galls, crevices in bark tree or wall, soil, water to household furnishings and human beds. The variety of diverse acarine parasites of insects, birds, reptiles and mammals sometimes make one wonders about their adaptability to survive in mammalian hair follicle, eyelids, ear canals to fur and feathers of animals and birds.

On one hand, phytophagous mites as pests of field crops and mites associated with granaries and warehouses cause heavy financial losses, other species have attained prominence as household pests and causative agents of allergic reactions in human beings and other mammals. Many parasitic forms act as vectors of disease causing organisms. With the appreciation of the probable importance of microarthropods in maintenance of soil fertility, amongst which mites constitute a dominant fauna, ecologists are directing their attention to this group of animals. Acari are among the major sources for biological control of various classes of pests including insects, nematodes and weeds.

The growth in the knowledge of all branches of acarology *viz.*, taxonomy, morphology, embryology, ecology etc. has been slow but steady and remarkable which led to the increase in acarological literature (Baker and Wharton 1952; Ghai, 1964; Evans, 1969; 1992; Grandjean, 1972-76; Jeppson *et al.*, 1975; Rodriguez, 1978; Channabasavanna and Virakta math, 1988, Gupta, 1985; 1986; Sadana, 1985; Woolley, 1988; Schuster and Murphy, 1991). Acarines have come a long way from the mere mention of 'small little waxies' by Aristotle to causing damage upto 80 per cent in agricultural crops and causative agents of allergy in more than 75 per cent persons associated with farm produce and their storage. Concurrent with the development of Entomology and Nematology, Acarology is also steadily gaining the status of a branch of plant protection science.

1.2 Relationship with Other Arachnids

The earliest known fossil, *Protacarus crani* Hirst although belonged to mid–Paleozoic era, the major adaptive breakthrough occurred in late Mesozoic and early Cenozoic times. Difference of opinion exists on the evolution of acarine radiation as Sharov (1966) put forward the concept of Acari evolution from pedipalpal ancestral stock during the middle Devonian time; Grandjean (1935) and others believed it to be of diphyletic origin with the Acariformes and Parasitiformes branching as separate entities from primitive arachnid stock.

The Chelicerates of class Arachnida are primarily terrestrial amongst which Acarina is the only subclass that contained group of specialized families which feeds on plants and cause deformities in different crops due to their feeding behaviour. They possess simple eyes and two pairs of mouthparts namely chelicerae and pedipalp which are adapted for grasping, piercing, sucking and tearing. Specialized pedipalpal or chericeral glands are present in some of the orders (Figure 1.2) while others have terminal glands. Like their silk secreting relatives Araneae (abdominal spinnerts) and Pseudoscorpionida (cheliceral glands), tetranychid (phytophagous) mites secrete silk from their palps.

In arachnids, body is divided into prosoma (six segments) bearing mouthparts and legs and opisthosoma (twelve segments) whereas in acarines distinct separation between prosoma and opisthosoma do not exist. However, the tagmosis is based on gnathosoma bearing mouthparts and idiosoma bearing legs and rest of the body. Different acarologists have used different terminology to express the varied tagmata in mites, which are given in Figure 1.1. Secondly, somatic segmentation is generally inconspicuous or absent in acarines whereas abdominal segmentation is a primary attribute in all the other subclasses except Araneae.

Acari show a wide variety of adaptations regarding life cycle and feeding habits. The majority of the mites are oviparous although some mesostigmatid and prostigmatid mites exhibit ovoviviparity and some parasitic forms exhibit viviparity. Regarding food habits, they are fluid feeders and either feed on liquid food or render their food fluid by secreting enzymes from the salivary glands. Once liquefied, the food is imbibed by the action of the suctorial pharynx.

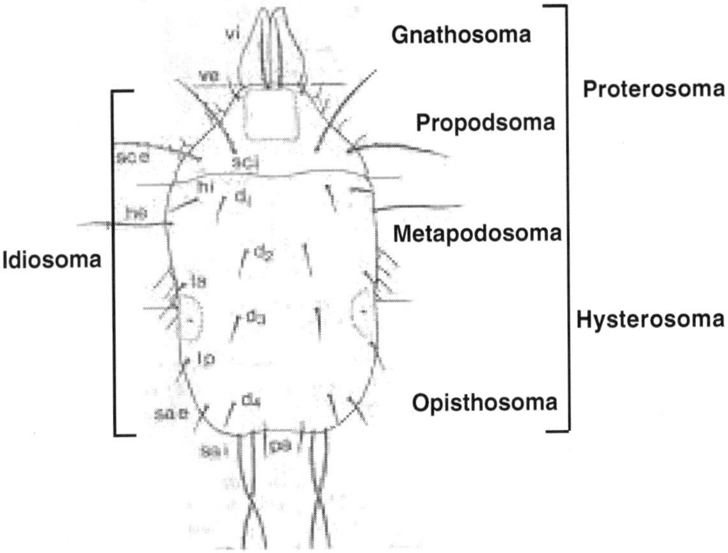

Figure 1.1: Terminology to Describe Acarine Tagmosis (Hughes, 1976)

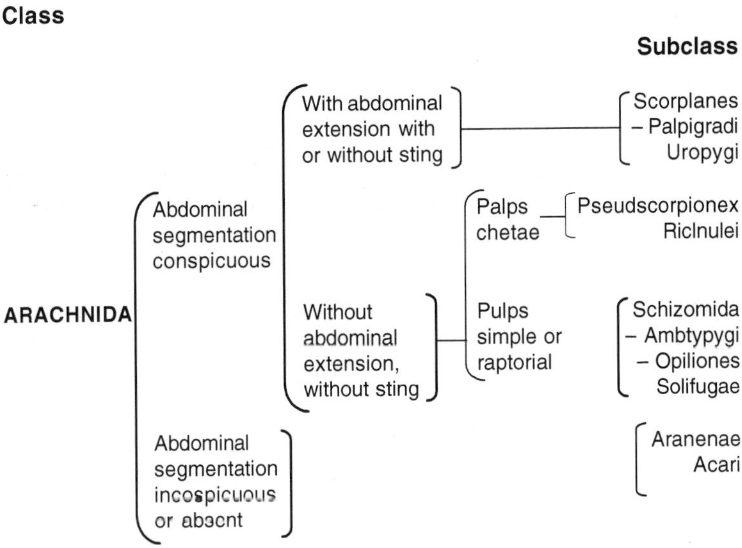

Figure 1.2: The Key to the Subfamilies of Arachnida (*Source*: Mohanasundaram, 2001)

1.3 Keys to Orders and Suborders

The suborders were defined on the basis of characteristics of the respiratory system by Evans *et al.* (1961) as evidenced by the taxa Tetrasitgmata (Holothyrida), Notostigmata (Opiloacarida), Metastigmata (Ixodida), Mesostigmata (Gamasida), Prostigmata (Actineida), Astigmata (Acaridida) and Cryptostigmata (Oribatida). Terminology used in the parentheses was proposed by Krantz (1978). The first four suborders were kept in order Parasitiformes and the last three in order Acariformes of subclass Acarina.

1.3.1 Order Acariformes

Lateroventral hysterosomal stigmata absent or inconspicuous, sensilla present or absent.

1.3.1.1 ASTIGMATA

Stigmata absent, sclerotization of adult integument weak or lacking, body whitish, size 200 to 1,200 μ, gnathosoma is not in camerostome, chelicerae strong, chelate and dentate, pedipalps two segmented and close to the side of infracapitulum, no sensory setae on propodosoma, epimera on I to IV legs normally well developed, the genital orifice, normally a longitudinal slit but occasionally transverse and located in the intercoxal region, it is often provided with two pairs of genital suckers in the female. This order includes stored grain pests, house dust mites, fungivores, detritus feeding mites and parasites of vertebrates.

1.3.1.2 CRYPTOSTIGMATA

Stigmata are always difficult to see (hence the name of the suborder), sclerotization of adult integument strongly developed, size 200 to 1,500 μ, body brown or black, gnathosoma with conspicuous rutella and lies in the camerostome, chelicerae strong, chelate and dentate, pedipalps usually with five free segments, sensory setae on propodosoma, the genital aperture in opisthosomal region protected by a pair of valves, usually three pairs of genital suckers. Members of this order are generally abundant in the upper layers of forest soils. They are generally algivores, fungivores or feed on decaying leaves and woods of higher plants.

1.3.1.3 PROSTIGMATA

Stigmata present and opening either between the bases of the chelicerae or on the anterior portion of the propodosoma, most heterogenous order, size varies between 100 to 16,000 μ, gnathosoma without rutella, well developed appendages, chelicerae usually hook like or styletiform and rarely chelate and dentate, pedipalps variable may be sensory appendage or modified for raptorial purpose, tibia with a strong claw or /and thumb claw complex, legs normally ends with a paired claw and tenent hairs. This order includes parasites of vertebrates, invertebrates and phytophagous, predatory species.

1.3.2 Order Parasitiformes

Conspicuous lateroventral hysterosomal stigmata present (1-4), sensilla absent. Notostigmata and Tetrastigmata are considered as rare and exotic suborders.

1.3.2.1 METASTIGMATA

Stigma with oval, rounded, comma shaped or subtriangulat stigmal plates, located between coxae III-IV or behind coxae IV, largest forms of acari, size sometimes exceeding 3 cm. in length (females of *Ornithodoros acinus*), Haller's organ on tarsus I, idiosoma dorsoventrally flattened and is sac like or sub circular in outline, gnathosoma with elongated hypostome armed with recurved teeth, external parasites (ticks) of vertebrates.

1.3.2.2 MESOSTIGMATA

Stigma located between coxae II and III or III and IV and sometimes surrounded by an elongate peritremal shield, peritreme long pipe like which start from stigmata and go anterior towards gnathosoma small non-leathery animals, with or without sclerotized plates, size 200 to 2,000 μ, smaller forms are pale and weakly sclerotized but generally idiosoma is partially covered by a number of chestnut brown shields, gnathosoma lies in camerostome, the ventral side of gnathosoma carries a maximum of four pairs of setae, the genital orifice is in the form of transverse slit in the inter coxal region, chelicerae three segmented, chelate, dentate but considerable modifications are found in parasitic species. Majority of mesostigmatids are free living in soil and decaying organic matter while others are adapted structurally and biologically as parasites of vertebrates and invertebrates.

Families can be identified on the basis of mouthparts; presence or absence of different shields; shape of peritremes; leg segmentation and presence or absence of empodium or claw. Chaetotaxy, poroidotoxy and adenotoxy are used widely in classification and identification of acarines up to species level.

1.4 Acarines: Their Types and Habitats

The Acari, which appear to be the largest biologically diverse group of Arachnida, vie for supremacy over insects in adaptability to any conceivable habitat although they are closer to spiders in phylogeny. These minute, cryptic creatures are widely distributed from inconceivable to conceivable locations. Their domicile ranges from earth to water to air and from sea to mountain peak. Their diversity is reflected in their structural and life history adaptations. Economics in agriculture is constantly affected by phytophagous acarines that infest variety of agricultural crops including horticultural, vegetable, ornamental, feed and fodder crops. Some species suck the cellular material of leaves (tetranychids), others make homes in the form of galls and cause severe deformities and also act as vector for many viral diseases (eriophyids). Many species dwell in crevices in bark, tree holes, forks of larger tree branches or in galleries of wood boring beetles.

Domestic mites (stored and housedust mites) live in close association with mankind. Stored mites live on the germ of many stored grains, thereby damaging the material both quantitatively and qualitatively. Housedust mites are our bedfellows, residing in the dust and lint of mattresses, pillows, an upholstered couch or sofa, carpets while feeding on the dander of human skin. Moreover, domestic mites are causative agents of allergic reactions in human beings.

Human and animal health is also adversely affected by transmission of disease and the irritation of bites caused by housedust mites and parasitic mites. The skin, eyelids, ears, nasal and respiratory cavities of vertebrates have special acarine inhabitants. A variety of species are habitants of insects like bees, beetles, cockroaches, grasshoppers and wasps. Invasion by acarines whether honeybee mites, house dust mites or some tarsonemids cause abandonment by the original tenants, leaving both house and pantry available to their new owners (acarines). Death of hosts also occurs, in severe infestation cases.

Examples of mites associated with soil are numerous. Some mites feed on the litter and are responsible for increased fertility of soil while others feed on insect pupae and nematodes and act as efficient biocontrol agents. Mites are also encountered in marine and fresh water, cold lakes, esturies and hot springs. Acarine biodiversity is discussed in Part I of this book under six chapters.

1.5 Acarological Techniques

It is often stated that mites make up in numbers what they lack in size. Therefore, mite specimen should be collected in quantity in order to obtain all sizes of the life cycle. Many entomological techniques have been used for collecting Acari but they are too small to be dealt with satisfactorily with these methods. But for standard microscopic procedures, they are too large in size. Consequently, many specialized techniques have been developed for mites housing different habitats. Techniques in relation to collection, preservation, preparation for study, culturing and bioassay are discussed in separate chapters in Part II.

1.6 Acarines as Bio-control Agents

The acarines are among the major, rather under utilized resources for biological control of many classes of pests including insects, nematodes and weeds. Biological control of pests by mites means any acarine activity (parasitism, predation, competition, disease transmission or any combination), which reduces pest population and the extent of their damage to below accepted economic injury levels. More than twenty families are associated with bio-control of other pests, however full potential is not estimated in majority of them. These are discussed in three chapters in Part III.

Esthetically, mites contribute a minute spectra of beauty and outstanding structure in terms of brightly coloured velvety skin of trombiculids, lacy integumental patterns on predatory mesostigmatids, delicate folding or fingerprinting of fur and feather parasitic mites and kinds of hairs and shields on their body.

In the end of this chapter, it may not be incorrect to say that mites have had impact in the past, affecting the present and may hold the key in future of rocket age.

PART I:
Acarine Biodiversity

Chapter 2

Plant Mites

Among the arachnids, Acari are the only group, which feeds on plants. The expanding nature of agriculture and changes in agricultural practices including increased usage of fertilizers, pesticides (Cohen and Gerson, 1989), irrigation and mono-cropping have aggravated the problem of destructive mite pests on different crops. Plant-feeding mites play an important role as agricultural pests of fruits, vegetables, forage crops, ornamentals and other agricultural crops. The damage caused by many species is still innocuous and often mistaken. Around 7,000 species of plant-feeding or phytophagous mites are known worldwide which occur in five families namely Tetranychidae, Tenuipalpidae, Tarsonemidae, Eriophyidae and Tuckerillidae. In India, a total of 660 mite species are known, out of which 169 (17 as major and 30 as minor pests) are represented by phytophagous mites. Crop wise status of mite pests in India is provided in tabular form for ready reference (Table 2.1). In various crops, 10-15 per cent losses are reported due to spider mites and in some cases, total loss is reported. In particular, 50-80 per cent in mango, 10-15 per cent in rice, 15-20 per cent in tea, 10-25 per cent in sugarcane, 13-31 per cent in brinjal, 25 per cent in okra, 27-39 per cent in chillies due to phytophagous mites are reported (Table 2.2). Severe infestation results in decrease in fruit size, plant vigour and total loss of crop. Some genera are

Table 2.1: Crop-wise Status of Mite Pests in India

Crops	*Total Mite Species*	*Total Mite Pest*	*Major Pest*	*Minor Pest*
Vegetables	**54**	**6**	**4**	**2**
All vegetables			*Tetranychus urticae*	
Brinjal			*T. neocaledonicus*	
Cowpea			*T. ludeni*	
Chilli, potato, tomato			*Polyphagotarsonemus latus*	
Cucurbits				*T. macfarlanei*
Fruit trees	**122**	**24**	**6**	**18**
Apple			*T. urticae, Panonychus ulmi*	
Citrus			*Eutetranychus orientalis, Brevipalpus phoenicis*	*Schizotetranychus hindustanicus, P. citri, B. californicus, Phyllocoptruta oleivora*
Guava, Pear			*B. phoenicis*	*C. kenyae*
Mango			*Aceria mangiferae*	*Oligonychus mangiferus*
Litchi			*A. litchii*	*O. mangiferus, O. beharensis*
Fig				*A. ficus, Eotetranychus hirsti*
Pear				*O. obovatus*
Banana				*O. indicus*

Contd...

Table 2.1–Contd...

Crops	*Total Mite Species*	*Total Mite Pest*	*Major Pest*	*Minor Pest*
Pomegranate				*O. punicae*
Grapevines				*O. mangiferus*
Loquat				*O. beharensis*
Ber				*Larvacarus transitans, Eriophyes cernuus*
Date palm				*Raoiella indica*
Cereals	**13**	**5**	**1**	**4**
Wheat			*Petrobia latens*	
Paddy				*O. indicus, O. oryzae, S. andropogoni, S. spinki*
Pulses	**13**	**4**	**1**	**3**
Red gram			*A. cajani*	*S. cajani*
Black gram, green gram				*T. urticae, P. latus*
Oilseeds	**9**	**3**	**2**	**1**
Castor, soybean			*T. urticae*	
Coconut			*A. guerreronis*	R. indica

Contd...

Table 2.1–Contd...

Crops	*Total Mite Species*	*Total Mite Pest*	*Major Pest*	*Minor Pest*
Ornamentals	**53**	**7**	**3**	**4**
Rose			*T. urticae*	*E.orientalis,B. phoenicis*
Zinia			*T. neocaledonicus*	
Marigold			*P. latus*	*B. californicus*
Jasmine				*A. jasmini*
Fibre crops	**8**	**3**	**2**	**1**
Jute			*P. latus*	
Cotton			*T. urticae*	*T. macfarlanei*
Plantation crops	**15**	**9**	**5**	**4**
Tea			*O. coffeae, B. phoenicis, P. latus, Acaphylla theae, Calacarus carinatus*	*T. urticae, B. obovatus*
Arecanut				*O. indicus, R. indica*
Commercial crops	**10**	**4**	**1**	**3**
Sugarcane			*O. indicus*	*A. sacchari, S. andropogoni, S. bancrofti*

Contd...

Table 2.1–Contd...

Crops	*Total Mite Species*	*Total Mite Pest*	*Major Pest*	*Minor Pest*
Fodder crops	**17**	**2**	**1**	**1**
Sorghum			*O. indicus*	
Grasses				*S. andropogoni*
Spices	**5**	**4**	**1**	**3**
Chilli			*P. latus*	
Coriander				*P. latens*
Cardamom				*B. phoenicis, Dolichotetranychus floridanus*
Shade trees	**19**	**2**	**1**	**1**
Neem			*E. orientalis*	
Saal				*O. beharensis*

polyphagous in nature and cause severe damage to large number of crop plants and trees.

Studies on the influence of weather factors show that high temperature coupled with low humidity is not only favourable to phytophagous mites but they cause relatively greater injury during this period (Andres, 1957). High temperature coupled with less rainfall also appeared to favour the increase in mite population (Lal, 1982; Puttaswamy and Channabasavanna, 1979). The nutritional status of plants affects the population growth rate of phytophagous mites (Wermelinger *et al.*, 1985). Walde (1991) further recorded that predators (Phytoseiids and Stigmaeids) were less effective in controlling *Panochynus ulmi* on trees with high nitrogen levels.

Table 2.2: Yield Losses Due to Different Phytophagous Mites

Mite Species	*Crop*	*Losses (per cent)*
Tetranychid mites	Vegetables	10-15
T.urticae	Okra	23-25
	Brinjal	13-31
	Cotton	20-30
	Pointed gourd	36
O. coffeae	Tea	5-11
S. andropogoni	Sugarcane	20-30
O. indicus	Sorghum	56
O. oryzae	Paddy	20-25
P. latus	Chillies	27-39
L. transitans	Ber	20-40
A. litchii	Litchi	30
A. mangiferae	Mango	50-80
A. cajani	Pigeon pea	15-30

In the life cycle, diapause is a widespread phenomenon in tetranychid, eriophid and phytoseiid mites. Biophysical and physiological mechanisms are involved in its initiation, maintainence and termination. Temperature, food, moisture and photoperiod have been implicated in diapause induction out of which photoperiod is considered to be the most important (Veerman,

1977; 1994). Diapause is facultative in nature in the sense that not every individual of every generation enters diapause. It may occur in the egg, larva, nymph or female but for each species the suppressed development normally occurs with in a particular stage.

2.1 Tetranychidae

These mites are generally referred to as spider mites. They are soft bodied, variously coloured, colony forming and many of these can spin webs to protect themselves from natural enemies and pesticides. There are approximately 106 species known out of which 19 species have taken a pest status. These mites possess long curved movable chelicerae. The fourth palpal segment bears a strong claw. The tarsi I and II and sometimes the tibiae bear specialized duplex setae. The claws possess tenent hairs and the empodium may or may not have tenent hairs. The female genitalia are the characteristic of the family. Normally, there are three pairs of propodosomal, four pairs of marginal, five pairs of dorsal and one pair of humeral setae.

2.1.1 Life History

There are five stages in the life cycle of these mites *i.e.* egg, larva, protonymph, deutonymph and adult stage. The nymphal and adult stages are initiated during the inactive period called protochrysalis, deutochrysalis and teliochrysalis. Mite fixes itself to a leaf or to its webbing during this period, legs are bent upon and a new cuticle is formed before casting off exuviae. They can develop from egg to adult in six to ten days. Egg laying by adult females can begin as soon as one or two days following maturity. Each female may lay up to 100 eggs in her 30 day lifespan. They occur on the underside of leaves, making coverage difficult. High fecundity and overlapping generations results in outbreaks and contributing to the development of insecticide resistance.

2.1.2 Nature of Damage

Both nymphs and adults feed on the leaf surface. White specks are formed on the leaves in later stages of infestation and general chlorosis occurs in patches. Tetranychid mites secrete certain substance into plant cells. Puncturing of new cells proceeds from one spot to another in the form of a circle, which results in the formation of small rounded chlorotic spots. Spider mites feed on leaf tissues and at the microscopic level it is estimated that roughly 50 per cent of the mass of an adult female spider mite is eaten per mite per hour. The numbers of photosynthetically active leaf cells that are punctured and emptied per mite, are 100 cells per minute. At the macroscopic level, damage from mite feeding can cause leaf bronzing, stippling or scorching. There is extensive webbing on leaf surface and black faecal dots are seen on the leaf surface.

Severe spider mite infestation cause major reductions in plant growth rates, flower formation and yield (Figures 2.1–2.2). Penetration of cells by mite stylets and injection of saliva cause both mechanical damage and changes in cell cytology and physiological and biochemical processes of non-punctured adjacent cells (Tomczyk and Kropczynska, 1985). In leaves damaged by mites, degeneration of chloroplast structure (Tanigoshi and Davis, 1978), reduction in stomatal (day time) transpiration and increase in cuticular (night time) transpiration (DeAngelis *et al.*, 1982) occur thereby reducing leaf gas exchange and inhibition of photosynthesis. Thus, in case of severe infestation, plants show yellowing and general drying of leaves, which drop prematurely. Damage symptoms peculiar to particular species is provided below to understand their behaviour.

2.1.3 Commonly Occurring Genera (Gupta, 2003)

Aponychus

12-13 setae on dorsum, genu 1, 2 and tibia 2 posses more than one setae.

Bryobia

Empodium without row of tenent hairs, rostrum with two pairs of prominent projections each with fan like seta, propodosoma with four fan like setae.

Eotetranychus

Empodium splits into tuft of hairs (3 pairs), two pair of para anal setae adjacent to one another.

Eutetranychus

Empodium without tenent hairs, tarsus I without duplex setae, empodial claw absent.

Oligonychus

Empodium without tenant hairs but empodial claw present with proximoventral hairs, one pair of para anal setae.

Panonychus

Empodium claw like with proximoventral hairs, without tenent hairs, two pairs of para anal setae, presence of dorsal body setae on tubercles.

Petrobia

Empodium with row of tenent hairs, rostrum without any projection or without fan like setae.

Schizotetranychus

Empodium claw like and splits into two parts, 2 pairs of para anal setae, dorsal body setae not on tubercles, duplex setae approximate to one another.

Tetranychus

Empodium without tenent hairs, presence of empodial claw and one pair of para anal setae, empodium splits distally into three pairs of hairs.

2.1.4 Damage Symptoms

Tetranychus and *Schizotetranychus* sp. prefer lower leaf surface.

Eutertanychus and *Oligonychus* sp. prefer upper leaf surface.

T. urticae

Chlorotic patches on leaves which turn brownish and drop off. Blackening of leaves, decrease in fruit size and yield is observed in pear.

T. ludeni

Yellow-bronze leaves in beans. In mulberry, white specks are formed which are enlarged to form large patches and give rusty, dry look. It is also a vector of Dolichos Enation Mosaic Virus.

E. orientalis

Infestation starts along the midrib and later spread to lateral veins; chlorotic patches also appeared due to feeding.

E. hirsti

Trasparent green patches on the leaves' under surface are observed initially which turn yellowish green, brown with rough and dry texture.

O. coffeae

Infested leaves turn brown and dry up.

S. andropogoni

White patches on sugarcane leaves are observed which turn brown and leaf dries.

P. citri

Stippling and light coloured spots on foliage which give greyish or silvery appearance.

P. ulmi

Normally found on upper surface of foliage, heavy infestation leads to reduction in fruit size and yield.

P. latens

L:eaves dry up from tip downwards, start showing yellowing appearance which ultimately dry out. Heavily infested plant gives sickly appearance.

2.2 Eriophyidae

These are small acarine agriculture pest with worm like annulated body. More than 500 species are known with 13 as pests of various crops. They measure from 150µ to 350µ in length and 35µ to 45µ in thickness. They are whitish or brownish in colour and possess two pairs of anteriorly placed legs. The body is divided into gnathosoma (rostrum), propodosoma (cephalothorax), and hysterosoma (abdomen). Rostrum is curved downwards and ocelli are absent. Sexes can be differentiated on the basis of genitalia.

2.2.1 Life History

There is no copulation in these mites. The male leaves the sperms in the form of spermatophores on the leaf surface. While moving on the leaf surface, the female genitalia comes in contact with spermatophores and are thus picked up by the females. Eggs are transparent and globular. After hatching, first stage nymph is formed with micro tubercles not fully developed on the body. It moults into the second stage nymph, which in turn moults into adult. It has got two pair of legs, fully developed micro tubercles, cephalothoracic shield pattern and genitalia.

Figure 2.1: Damage Caused by Tetranychid Mites

2.2.2 Nature of Damage

The mites occur on all parts of a plant and may or may not exhibit the symptoms of damage. Based on type of injury (Figure 2.3), they have been classified as under:

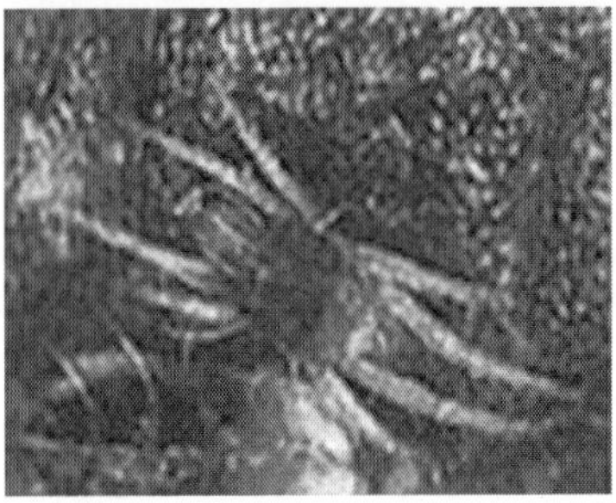

Panonychus ulmi

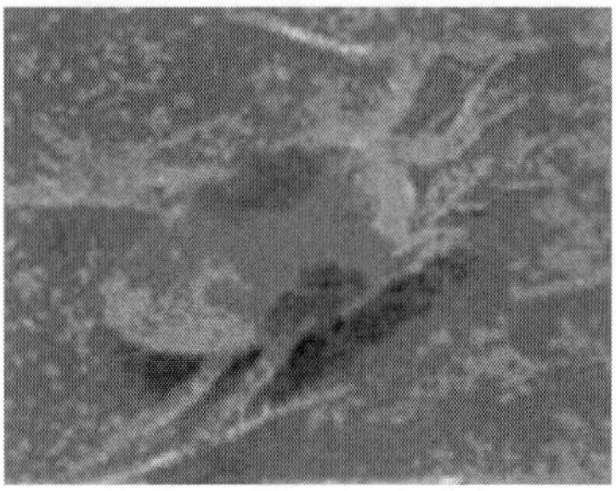

Tetranychus urticae

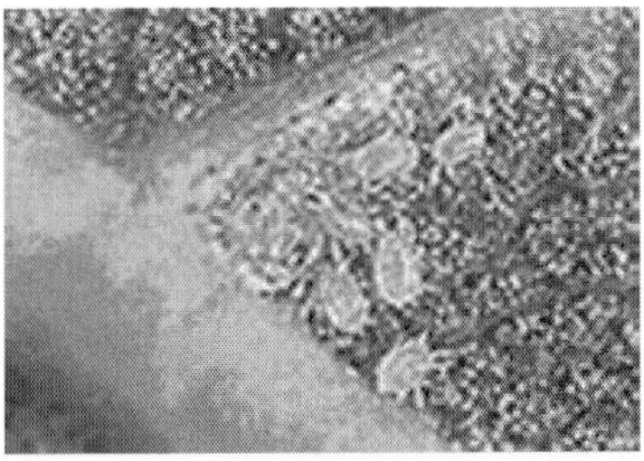

Tydeid Mites

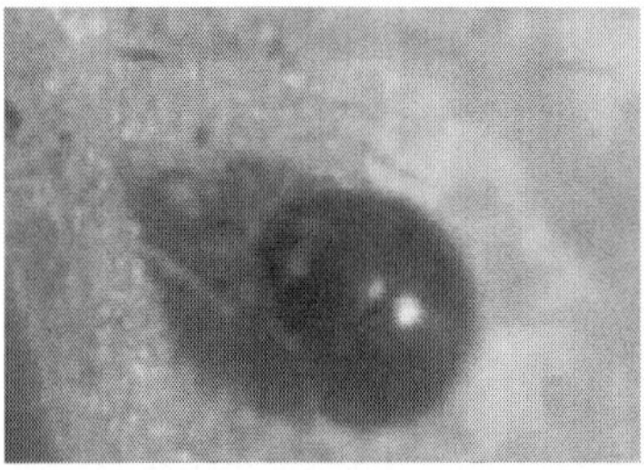

Beetle Feeding on *T. urticae*

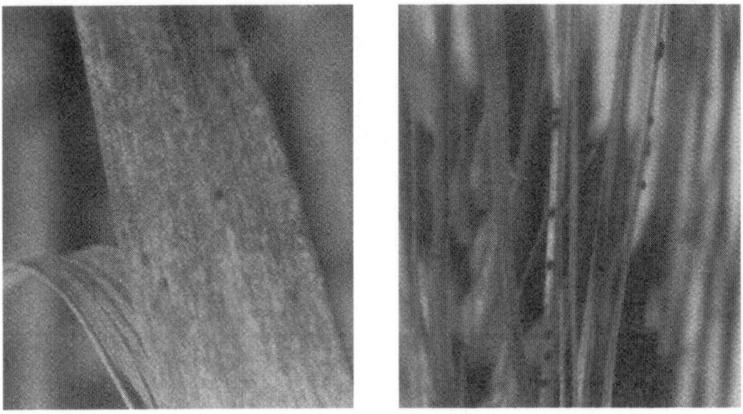

Damage Caused by *Petrobia latens* (Sharma and Srinivasa, 2004)

Figure 2.2: Tetranychid Mites and their Damage

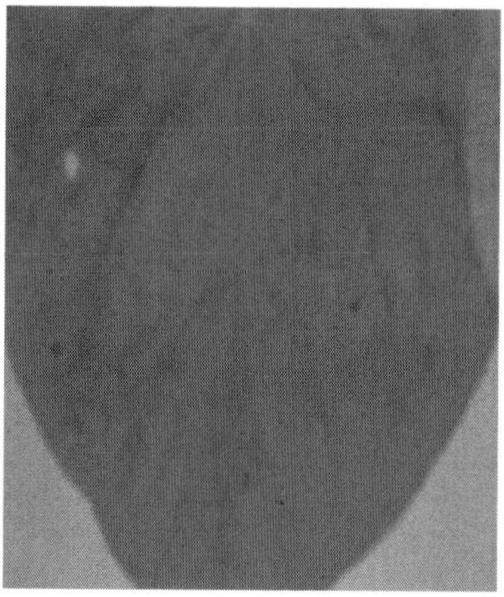

Dhooria *et al.*, 2005

Srinivasa *et al.*, 2004

Dhooria *et al.*, 2005

Figure 2.3: Damage Caused by Eriopyid Mites

Gall Formers

Due to feeding on various plant parts hypertrophy of cells occur. It results into formation of galls on leaves, flower buds and stem. Different types of galls like pouch galls (*Pongamia* sp.), bead galls (*Ficus* sp.), finger galls (*Pongamia* sp.) have been described.

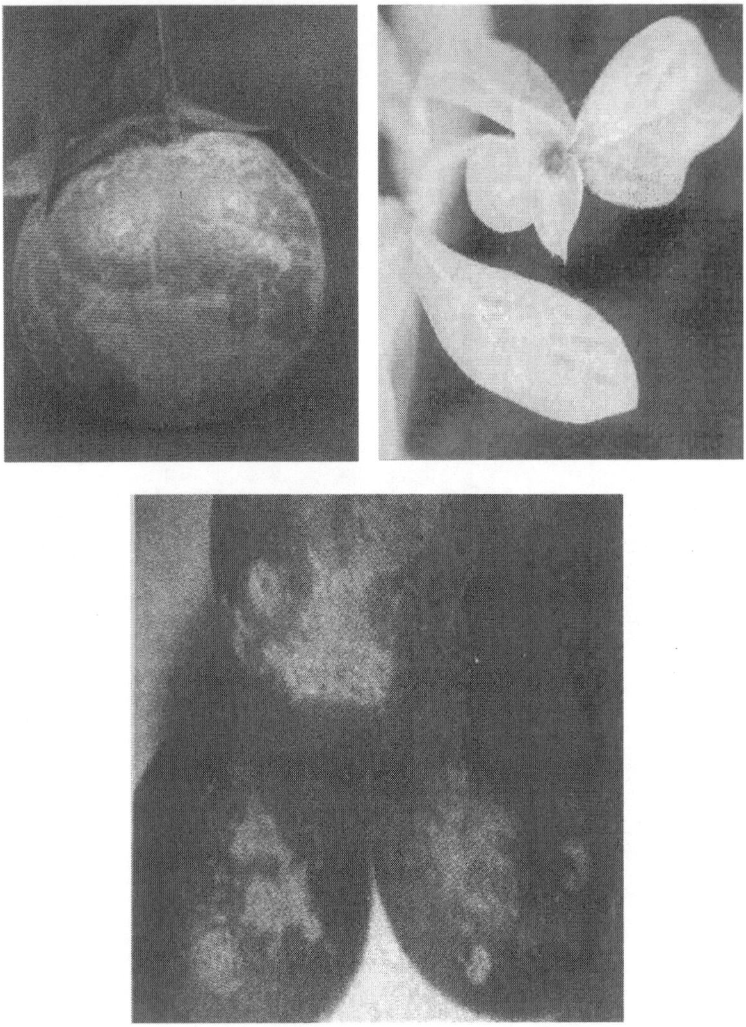

Figure 2.4: Damage Caused by Tenuipalpid Mite (Dhooria *et al.*, 2005)

Bud Mites

Some species of eriophyids feed on developing vegetative buds within unopened leaves, as a result, these buds get killed.

Leaf Rollers

These mites roll the whole leaves or only edges of leaves and feed within the rolls.

Erineum Formers

Due to feeding by mites, epidermal layer of cells produce hair like out growths, which produce the erineum.

Blister Mites

Some species cause formation of blisters on the leaf sheath and feed within.

Vagrants

These mites may not show any symptoms except rusting of leaves. Such mites are found on both surfaces of leaves.

Apart from these injuries, some species play a vital role in virus transmission like Pigeonpea Sterility Mosaic Disease by *Aceria cajani*, Wheat Streak Mosaic Disease by *A. tulipae*, Sugarcane Streak Mosaic Virus by *A. sacchari*, Fig Mosaic Virus by *A. ficus* etc.

2.2.3 Commonly Occurring Genera (Gupta, 2003)

Acaphylla

Body fusiform with distinct tibia, feather claw divided.

Aceria

Worm like elongated body, tibia distinct and dorsal shield seta are pointing backwards.

Calacarus

Body fusiform with distinct tibia, feather claw simple with central stem undivided, robust, deep purplish body, secretes white wax on shield and form five longitudinal dorsal ridges.

Cisaberoptus

Tibia is reduced or completely fused with tarsi.

Eriophyes

Worm like elongated body with distinct tibia, dorsal shield seta are pointing forward.

Phyllocoptruta

Flattened, yellow, fusiform body, feather claw simple with central stem undivided, anterior shield lobe project well over rostrum.

2.2.4 Damage Symptoms

A. litchi

Infestation leads to deformation, curling of leaves, chocolate brown erineum on leaf's lower surface, leaves become dry and fall.

A. lycopersici

Curling of leaves, appearance of silver gloss on lower surface, roughened appearance on leaves of brinjal, tomato. There is loss of hair, change of colour and cracks appear in stems.

A. mangiferae

Mango malformation, suck sap from buds which cause necrosis of tender tissues.

A. guerreronis

Injury in floral tracts, injured area become shrivelled, brown and cracks appeared in fruits.

A. cajani

Diseased plant show dwarfing of leaves, bushy habitat, yellowish green leaves, complete suppression of flowering and fruits.

A. jasmini

Cause red blisters on the inner surface of leaves, malformed floral parts.

A. sacchari

Formation of blisters of various shapes on inner surface of sugarcane leaf sheath. The colours of blisters are green then rusty.

E. sheldoni

Lemon bud scales are blackened, multiple budding on infested twigs, abnormal floral parts. In orange, fruits are flattened vertically.

P. oleivora

Damaged fruits become silvery, reddish brown or purplish black, affected fruits have thicker skin and show rust spots.

Acaphylla theae

Veins and margins show pinkish discolouration, infested leaves are pale and leathery.

Calacarus carinatus

Infested leaves are initially in copperish brown colour, which turns purplish bronze.

Cisaberoptus kenyae

White silvery coating on upper surface of leaves under which mites live.

2.3 Tenuipalpidae

They are also called as false spider mites as some species resemble tetranychid mites but they cannot spin webs. They are reddish in colour, dorso ventrally flattened, colony-forming mites who generally feed on lower surface of leaves, near the midrib or veins. Some species form galls and feed within while others feed on bark of trees, leaf sheaths and floral heads. Six species out of 98 species known, has taken a status of pest in agricultural crops

There are three pairs of propodosomal setae, however, the number of setae on hysterosoma may differ. Humeral setae are always present along with 5-7 pairs of other dorso lateral setae. A series of dorso sublateral setae are present, 1-4 pairs may occur between the lateral and dorso central setae, which usually occur in a longitudinal line. Chelicera is long and whip like. The male genitalia consist of terminal stylets. Aedaegus is long and tapering while the female genitalia consists of a simple platelet usually with 2 pairs of posterior setae. There may be a ventral platelet anterior to the genetic plate, both of which have generic significance.

2.3.1 Life History

Female lays eggs along the midrib on the lower surface of leaves. Eggs are rectangular with striations on the chorion. Eggs hatch into larvae, which bear three pair of legs. Larvae moult into protonymph with 4 pair of legs and then to deutonymph. All these three stages are stationary while adults move slowly.

2.3.2 Nature of Damage

These mites generally feed on the ventral suface of leaves near the midrib or veins. There is bronzing and rusting symptoms on the lower surface of leaves due to feeding of nymphs and adults (Figure 2.4). Some species form galls on the leaves and stems of plants. Each species has peculiar damage symptoms in host plant, which is provided here for better understanding of their behaviour.

2.3.3 Commonly Occurring Genera (Gupta, 2003)

Brevipalpus

Pear shaped, oval or round body, nymphs and adults with four pairs of legs, pedipalp four segmented and short body setae.

Dolichotetrancyhus

Body oval, pear shaped or round in shape, nymphs and adults with four pair of legs, pedipalp three segmented, flagellate setae on posterior margin absent.

Larvacarus

Body oval, elongated, nymphs and adults with three pairs of legs.

Raoiella

Body round, nymphs and adults with four pairs of legs, pedipalp two segmented, flagellate setae on posterior margin absent, some of the body setae are long, club shaped.

Tenuipalpus

Pear shaped, oval or round body, nymphs and adults with four pairs of legs, pedipalp three segmented, anterior portion of body broadened and distinguished from narrow posterior margin, long, flagellate seta on hysterosoma.

2.3.4 Damage Symptoms

There is bronzing and rusting symptoms on the lower surface of leaves due to feeding of nymphs and adults.

Raoiella indica

Showed reddish spots on leaves.

B. californicus

Infested leaves produce brownish patches.

B. phoenicis

Infested leaves turned pale yellow and develop brownish patches.

B. obovatus

Produced chlorotic patches with concentric rings of reddish resinous material.

D. floridanus

Feeding is at basal parts of leaves which cause rust like symptoms.

L. transitans

Mites form galls on ber.

2.4 Tarsonemidae

This family includes phytophagous, fungivorous and parasitic mites of scale insects and honeybees. These mites are very small in size varying from 100-300 μ in length. Body is soft, round, oval, white and glossy in colour. The anterior pair of legs bears more dense setae than the body and posterior legs. It also bears special sensory setae. These tiny mites are fast moving.

Sexual dimorphism is distinct. Female is provided with convex dorsum. Males are small in size than females with different body contour. Mouthparts are contained in capsular head called the capitulum. Mouthpart consists of a pair of stout pedipalp and styliform-paired chelicerae.

Rest of the body comprises of idiosoma, which is transected by a main body structure. The unsegmented part, which is anterior to the suture, is called propodosoma, which may be prolonged anteriorly to form the cephalothoracic shield or hood. The portion of the body behind the suture is called hysterosoma, which may further be divided into anterior portion called the metapodosoma, the portion from body suture to hind margins of coxae IV and the posterior portion behind the legs is called opisthosoma.

2.4.1 Life History

The female lays eggs singly. They are white, oval in shape. Eggs hatch into larvae, which is opaque white in colour and has three pair of legs. The anterior pair of legs is situated as in adults while the posterior pair of legs as legs III of the adults. Opisthosoma has a peculiar enlargement in the form of a triangular plate like structure, which is more distinct in males. After completion of larval stage, mites enter into quiescent pupal stage and transforms into the adult.

2.4.2 Nature of Damage

They usually infest the tender portion of plants and suck the sap from buds, leaves, shoots, flowers and stem sheath. They cause curling, crinkling and brittleness of foliage but shows little leaf symptoms. The injury caused by this group is often mistaken as a disease symptoms caused by pathogenic microorganisms. Damage symptoms of the common genera are given below in different host plants.

2.4.3. Commonly Occurring Genera (Gupta, 2003)

Polyphagotarsonemus

Terminal claw of fourth leg very much-reduced in male, in female leg IV ends in a pair of whip like setae, femur IV without any flange like projection.

Steneotarsonemus

In male, terminal claw well developed, curved and femur bulged inward into a flange like projection, leg IV of female ends into a pair of whip like setae.

Hemitarsonemus

Gnathosoma circular, pedipalp and chelicerae minute

2.4.4 Damage Symptoms

They usually infest the tender portion of plants and suck the sap from buds, leaves, shoots, flowers and stem sheath. They cause curling, crinkling and brittleness of foliage but shows little leaf symptoms.

S. spinski

Increase in percentage of empty grains. They carry spores of rice sheath rot fungus, which causes brown spots on rice sheath and grains. This disease is known as 'Sterile grain Syndrome'

S. bancrofti

Internodes give scabby corroded appearance, transparent depressions on young stalks.

P. latus

In chillies, it feeds on lower surface of leaves causing leaves to become rigid and curled, prevents flower and fruit development. This disease is termed as 'Murda disease'. In citrus, there is bronzing, roughening and crinkling of leaves. In potatoes, oily blackish spots appear on the under surface of young leaves which turn reddish. From tip the plant wither, auxillary buds are killed. This disease is known as 'Tambera disease'.

H. latus

Leaf remain smaller in size, crumpled leaves are observed which turn brownish.

2.5 Tuckrellidae

This is the smallest phytophagous family, which includes four species. These are brightly coloured having fan like dorsal body setae and long whip like caudal setae. *Tuckerella* sp. is most common genus occurring in sporadic form in various host plants. The mites do not have much importance, as they do not cause any economic damage.

2.6 Management Strategies for Phytophagous Mites

Introduction of broad spectrum pesticides for general pest control, use of high yielding varieties, development of resistance among mites to various acaricides have made it difficult to control phytophagous mites. Cultivation of resistant varieties, cultural control, mechanical control, biological control and need based application of acaricides are considered to be practical solutions for management of plant mites. Some of the management strategies are discussed under although specific control measures are also given under different phytophagous mite descriptions.

2.6.1. Cultural Control

Removal and destruction of damaged leaves and plant parts, removal of ratoon crops, weeds etc., avoiding monoculture, adjusting the date of sowing, inter cropping, need based irrigation and fertilizer application are some of the important practices which minimize the mite population in various crops. In coconut, removals of infected part, avoidance of excessive irrigation and fertilizer significantly reduced the *Aceria guerreronis* (Keifer) population (Alenear *et al.*, 1999). Pruning technique is also effective in checking and reducing the pest mite problem. *Brevipalpus phoenicis* occurs throughout the year on tea leaves with peak population in May-August around petioles. Pruning the tea leaves during this peak season significantly reduce the pest population (Gope and Das, 1992). Likewise, pruning and burning of infested leaves and twigs of horticultural crops effectively reduced the *Panonychus ulmi*, *Aculus schlechtendali* in apple (Koschier, 1997) and *A. litchi* in litchi (Prasad and Singh, 1981). In some cases, higher predatory population was recorded after pruning (Graften *et al.*, 1995), which reduces the pest population.

Intercropping with host or non-host crop is also done to reduce the pest population in target crop. Scanty literature is available against various pest mites. Cowpea intercropped with cotton (2:1) reduces the *Tetranychus arabicus* population on cotton (Omar *et al.*, 1993); beans intercropped with maize/onion/garlic showed reduced *T. urticae* damage. Planting of *Azeratus conyzoides* as a ground cover in hilly citrus orchards increased the summer relative humidity and decreased the temperature thereby improving the conditions for natural enemies. These plants provide alternative food and oviposition site for predatory mites, which suppress the autumn and spring damage caused by *P. ulmi* (Zhou *et al.*, 1994).

2.6.2. Soil Sterilization

Studies on sterilization techniques are available in bulb mites in gladiolis and onion. Sterilization of loosened soil with steams produced by soil sterilizer at 60°C for 30 minutes or 80°C for 20 minutes resulted in nearly 100 per cent mortality of bulb mites (Liu *et al.*, 1998). Bromide fumigation also gave effective control.

2.6.3. Host Plant Resistance

Differences in population increase of mites on different species, varieties or genotypes of host plants have been demonstrated by a number of workers (Table 2.3). The importance of leaf surface texture and contour in influencing the reproductive potential of tetranychid mites has long been recognized. Such features serve as good substrate for attachment for webbing or provide direct protection from harmful micro-weather on leaf surface. Strong webbers like *Tetranychus* spp. prefer rough surfaces to smooth ones. Leaf hairs could very well serves as substrate for attachment of webbing. Likewise, sugarcane varieties with spinous out growth on the leaves (Khanna and Ramanathan, 1947) were susceptible to *Oligonychus indicus*. Resistance to mites was attributed to glandular hairs in tomato varieties (Stoner *et al.*, 1968), okra leafed glabours genotypes in cotton (Wilson, 1992) and fruits with tightly pressed sepals in coconut (Moore and Alexander, 1990).

Table 2.3: Important Resistant Varieties/Cultivars against Phytophagous Mites

Sl.No.	*Crop*	*Mite pest*	*Variety/Cultivar*	*Reference*
1.	Brinjal	*Polyphagotarsonemus latus*	RHR-58	Misra *et al.*, 1990
		Tetranychus telarius	Muthatakeshi, 'Banta'	Lall *et al.*, 1965
2.	Chillies	*P. latus*	Jwala	Naitam *et al.*, 1990
3.	Cotton	*T. urticae*	LAFR 3159, He-BR-S-LA, 3382-6-OSN Pima 5-2	Leigh *et al.*, 1984
4.	Maize	*T. urticae*	B-96	Guthrie *et al.*, 1991
5.	Okra	*T. cinnabarinus* *T. telarius*	EC-329390, IC –14-1065 Sanjam	Ramaraju *et al.*, 1997 Gulati, 2004
6.	Tomato	*T. cinnabarinus*	*Lycopersicon esculentum*, *L. pervianum*	Wolfenbarger, 1965 Stoner *et al.*, 1968 Gentile *et al.*, 1969
7.	Tea	*Oligonychus coffeae*	MT-18, TR-2027	Thirugnanasunthaun and Amarasinghe, 1990
8.	Coconut	*Eriophyes guerreronis*	Kenthali rounded fruits less susceptible	Moore and Alexander, 1990
9.	Sugarcane	*O. indicus*	Leaves without hairs	Khanna and Ramanathan, 1947

2.6.4. Predatory Mites as Biocontrol Agents

The predatory mites belong to the families Phytoseiidae, Ascidae, Anystidae, Bdellidae, Cheyletidae, Cunaxidae,

Erythraeidae, Raphignathidae, Stigmaeidae, Tydeidae, etc. (Tables 2.4, 2.5). Most of these are voracious feeders of phytophagous mites.

Table 2.4: Predatory Mites Associated with Phytophagous Mites

Predatory Mite	*Phytophagus Mite*
Phytoseidae	
Amblyseius aceriae	Mango bud mite
A. aerialis	*Panonychus citri*
A. alstoniae	*Tetranychus cinnabarinus, Eutetranychus orientalis, Oligonychus indicus*
A. assamensis	*T. cinnabarinus*
A. bhadrakaliensis	*T. cinnabarinus*
A. bindrai	*T. cinnabarinus*
A. channabassavanni	*Raoiella indica, T. fijiensis*
A. cocciniae	Tetranychids, *P. citri*
A. coccoscocius	*Brevipalpus phoenicis*
A. concordis	Tetranychids, tenuipalpids, eriophyids
A. cucumeris	*B. phoenicis, E. orientalis, O. mangiferus, T. cinnabarinus*
A. deleoni	*Acaphylla theae, Calacarus carinatus*
A. delhiensis	*E. orientalis, O. indicus, T. cinnabarinus Schizotetranychus andropogoni*
A. dohooriai	*O. indicus*
A. fallacis	*Oligonychus* sp., *S. andropogoni, T. cinnabarinus*
A. finlandicus	*E. orientalis, T. cinnabarinus*
A. herbicolus	*O. coffeae, B. phoenicis*
A. hibisci	*E. orientalis*
A. indicus	*Tyrophagus putrescentiae, Petrobia latens*
A. kalimpongensis	*Brevipalpus* sp., *Tetranychus* sp.
A. lablabi	*Tetranychus* sp.
A. largoensis	*P.citri, Eutetranychus* sp., *E. orientalis O. mangiferus, O. coffeae, B. deleone*
A. longispinosus	*T. ludeni, S. andropogoni*
A. mangiferae	Eriophyid mite
A. mcmurtryi	*T. cinnabarinus*

Contd...

Table 2.4–Contd...

Predatory Mite	Phytophagus Mite
A. multidentatus	*T. cinnabarinus, T. ludeni, O. indicus, B. phoenicis*
A. ovalis	*T. cinnabarinus, O. mangiferus, S.cajani*
A. paraerialis	*T. cinnabarinus*
A. potentiallae	Eriophyid mite, *T. cinnabarinus*
A. reticulates	*O. indicus*
A. rhabdus	*A. theae, C. carinatus*
A. rhododendronis	*T. ludeni*
A. sacchari	Eriophyid mite, *O. indicus*
A. salebrosus	*T. cinnabarinus*
A. sukaensis	*E. orientalis*
A. tetranychivorous	*T. cinnabarinus, T. ludeni, R. indica, E. orientalis, O. indicus, O. beharensis, Schizotetranychus* spp.
Indoseiulus ricini	*Tetranychus* sp.
Phytoseius coheni	*B. phoenicis*
P. fici	*B. californicus*
P. kapuri	*T. cinnabarinus, T. neocaledonicus*
P. minutus	*Tetranychus* sp.
P. persimilis	*R. indica, T. cinnabarinus, T. ludeni, T. fijiensis, T. neocaledonicus*
P. punjabensis	*Eotetranychus hirsti*
P. roseus	*E. orientalis, O. indicus*
Typhlodromus bambuscolus	*T. cinnabarinus*
T. caudiglans	*A. mangiferae*
T. channabasavanni	*B. phoenicis*
T. eharai	*O. coffeae*
T. fleschneri	*T. cinnabarinus*
T. gopali	*Eotetranychus* spp.
T. homalii	*Oligonychus* sp., *E. orientalis*
T. kodiakanalensis	*Tetranychus* sp.
T. rhenanus	*B. californicus*, Eriophyid mite, *E. orientalis*
T. rickeri	*Tetranychus* sp.
T. roshaniali	Eriophyid mite

Contd...

Table 2.4–Contd...

Predatory Mite	*Phytophagus Mite*
Cheyletidae	
Cheletogenes ornatus	*T. punicae, Larvacarus transitans*
Cheletomorpha lepidopterorum	Tetranychids
Cheletonella sp.	Tetranychids
Chelacaropsis sp.	Tetranychids
Cheyletus eruditus	*E. orientalis*
C. malaccensis	*Schizotetranychus* sp., *Brevipalpus* sp.
C. polymorphus	*T. cinnabarinus*
Hemicheyletia bakeri	*B. phoenicis, B. californicus, Tetranychus* sp.
Paracheyletia pyrilformis	grapevine mites
Stigmaeidae	
Agistemus fleschneri	*E. orientalis, B. obovatus, A. mangiferae, T. urticae*
A. floridanus	*T. cinnabarinus*
A. granulus	*E. orientalis*
A. macromatus	*E. orientalis*
A. miriabilis	*O. mangiferus*
A. industani	*E. orientalis, T. ludeni*
A. yunusi	*Tetranychus* sp.
A. herbarius	*O. mangiferus, T. urticae*
A. terminalis	*Brevipalpus* sp., *A. theae*
Zetzellia sp.	Tetranychids
Eryngiopus sp.	*T. cinnabarinus*
Anystidae	
Anystis baccarum	Tetranychids
A. indicus	*T. urticae*
Tencateia sp.	*E. orientalis, Brevipalpus* sp.
T. kanthiensis	*Tetranychus* sp.
Walzia sp.	Tetranychids
Cunaxidae	
Cunaxa setirostris	*T. cinnabarinus, E. orientalis, Oligonychus* sp., *O. mangiferus, S. andropogoni*
C. taurus	*Tetranychus* sp., *O. indicus*

Contd...

Table 2.4–Contd...

Predatory Mite	*Phytophagus Mite*
Bdellidae	
Bdellodes sp.	*O. indicus*
Spinibdella cronini	*B. californicus*
S. depresa	*O. mangiferus*
Ascidae	
Lasioseius mcgregori	*Tetranychus* spp., *O. indicus*
L. quadriselosus	*Tetranychus* sp.
Tydeidae	
Pronematus elongates	*E. orientalis*
P. fleschneri	*E. orientalis, T. neocaledonicus*
P. sextoni	*Polyphagotarsonemus latus*
Tydeus kochii	*B. califoricus*
Tydeus spp	*O. indicus*
Others	
Acaropsella volgini	*P. latens*
Bechsteinia schneieri	*B. phoenicis, Tetranychus* sp.
Leptus giganticus	Tetranychids

2.6.4.1 Phytoseiidae

This group has received maximum attention globally as they are reported as predators of phytophagous mites and small insects (Hoy and Smilanick, 1981; Yaninek and Herren, 1988). These mites are whitish/creamish/reddish or light brownish in colour, fast moving and abundant in nature. They have very long legs, which help them to run very fast. They have wide range of food habits from carnivores to non-animal food (pollen, honey, nectar, plant sap) eaters. Some prefer specific food to complete life cycle while others have flexible food habits. These mites have several advantages over other predatory mites because of high fecundity, abundant availability, good searching ability, dispersal rate, adaptability to different ecological niches and high degree of prey specificity. Among the 173 species that are so far known from India, 18 species have been found actively feeding on phytophagous mites.

Table 2.5: List of Predatory Mite Families with the Number of Species and Genera Associated with Plants

Sl.No.	*Predatory Families*	*No. of Species*	*No. of Genera*
Mesostigmata			
1.	Phytoseiidae	173	11
2.	Ascidae	9	5
Prostigmata			
3.	Anystidae	9	3
4.	Bdellidae	14	6
5.	Cheyletidae	15	10
6.	Cunaxidae	21	4
7.	Erythraeidae	16	7
8.	Eupodiadae	2	1
9.	Raphighathidae	3	2
10.	Stigmaeidae	25	6
11.	Tydeidae	21	5

Source: Gupta, 2003.

2.6.4.2 Stigmaeidae

These are probably next to phytoseiidae as far as predatory efficiency is concerned. These are yellowish/reddish/light brownish in colour, ovoid or elongated in shape, which occupy various habitats. These mites cannot run very fast hence these are used for controlling slow moving mites and for destruction of mite eggs. *Agistemus* spp. and *Eryngiopus coimbatorensis* are found to be of high promise as biocontrol agents, details of which are given in Table 2.4 and 2.6.

2.6.4.3 Anystidae

These are round, long legged, reddish and soft bodied mites. They are very fast moving and start making whirling movement as soon as touched. *Anystus* spp. and *Tencateia* spp. (Table 2.4) are reported from India but practically nothing is known about their biology and predatory eficicency.

2.6.4.4 Cunaxidae

These are reddish/yellowish/brownish mites, fairly fast moving and are known to be active predators of phytophagous mites. They have very strong and thorny mouthparts but their number is limited in nature. Some of the mites are known to fasten their preys with silken threads secreted by their mouthparts. So far, only one genus *Cunaxa* is reported from India having high predatory potentiality.

2.6.4.5 Bdellidae

These are fast moving, medium sized and reddish in colour. They prey upon phytophagous mites and small insects like collembola. So far, no species of high predatory potentiality has been noticed in field.

2.6.4.6 Cheyletidae

These are free living predators of phytophagous, stored and house dust mites. Comb like or sickle like setae are present on pedipalp tarsus. *Cheletogenes, Cheyletus, Hemicheyletia, Paracheyletia* are important species whose interaction with phytophagous mites are well documented in literature. *H. bakeri* (Ehara) feeds on spider mites and requires 22-25 days to complete its development. Feeding potential is 7-8 preys/day. Fecundity is 24 eggs per female but in suitable environmental conditions it can reach to 65 eggs per female. Laing (1973) indicated weak numerical response of *H. bakeri* to spider mite population fluctuations on strawberry leaves. The predator foraged on a limited area of the total distribution of prey, which limits its potential in field. Feeding potential of various predatory mites is known against different phytophagous mites (Table 2.6).

2.6.5. Insects and Spiders as Bioagents

A few insect species belonging to the order Coleoptera, Neuroptera, Hemiptera, Thysanoptera and Diptera are predatory on mite pests (Table 2.7). A single grub and adult of coccinellid beetle, *Scymnus gracilis* consume 41-72 eggs and 40-50 mites of *O. indicus* per day. *Stethorus* sp. is an effective predator of phytophagous mites. About eight species are reported to feed upon mites. The life cycles of *Stethorus* is completed in about 2 weeks and it requires 15 adults and 50-100 eggs or young mites to have sufficient food for egg production. In absence of mites, *Stethorus* feeds on raisins, nectar or

Table 2.6: Important Predatory Mites Alongwith their Feeding Rate, Development Period and Prey Stage Consumed

Sl.No.	*Predator*	*Prey Stage Consumed*	*Developmental Period*	*Fecundity*	*Feeding Rate*	*Reference*
1.	*Amblyseius vannai*	Larva, nymphs	85-107 h for ♀ 84-113 h for ♂	27.3 eggs (2.7±0.6/day)	14-19 eggs of *R. indica* by ♀ 15-38 eggs of *R. indica* by ♂	Daniel, 1981
2.	*A. alstoniae*	Nymphs, eggs of *E. orientalis*. All stages of *T. macfarlanei*, *B.phoenicis*			10.16 protonymph of *E.orientalis* by ♀ 46.26 adults of *T. macfarlanei* by ♂ 1:10 & 1:20 ratio effective for *B.phoenicis* control	Dhooria, 1980 Jose *et al.*, 1989 Kumari and Sadana, 1995ab
3.	*A. concordis*	Castor pollen preferred over tetranychids, tenuipalpids, eriophyids	120h at 20°C, 63 per cent RH 66.8 h at 35°C, 55.3 per cent RH			Nageshchandra *et al.*, 1999
4.	*A. finlandicus*	Nymphs, eggs of *E. orientalis*, *B. phoenicis*	Varied with temperature		When only *B. phoenicis* eggs wee fed, the progency was female only	Sharma and Sadana, 1989
5.	*A. ovalis*	Tetranychids	127 h on castor pollen 12.3 d	39.2±4.0 eggs 30.6 eggs	10.14 larvae/female/day 3.86 adult/female/day	Borah and Rai, 1989 Nangia and Channabasavanna, 1995

Contd...

Table 2.6–Contd...

Sl.No.	*Predator*	*Prey Stage Consumed*	*Developmental Period*	*Fecundity*	*Feeding Rate*	*Reference*
6.	*A. rhododendronis*	*T. ludeni*			Effective in checking *T. ludeni* population within 4 weeks of release	Krishnamoorthy, 1983
7.	*A. longispinosus*	Tetranychids	99.11 h for & 95.30 h for %		250 predators per 15 plants effectively controls	Nangia and Channabasavanna, 1995; Malik *et al.*, 1999
8.	*A. tetranychivorus*	*T. ludeni* stages eggs and nymphs of *R. indica*	140±7.5 h for & 136.2±6 h for % 142.28 h at 24°C, 65% RH	45.3±7.1 eggs 26.3 eggs on *R. indica*	26.74 preys/female 1:6 ration gave full control in 10 d in okra	Puttaswamy, 1978 Jagdish and Nageschandra, 1982 Channabasavanna, 1981 Nangia and Channabasavanna, 1995
9.	*Phytoseiulus persimilis*	Tetranychids	5.03 d at 25°C 3.84 d at 29°C	82.4 eggs	Release technique available in French bean	Krishnamorthy, 1988
10.	*Agistemus industani*	*T. ludeni* eggs		3 eggs/day	15 prey eggs/day	Rai and Singh, 1999
11.	*Cheletogenes ornatus*	*L. transitans*			5-60 mites in 12 h	Sharma and Kushwaha, 1983

honeydew or become cannibalistic. The beetles @ 400-500 per tree reduced the brown mite in Avacado. *Clanis sorews*, *Scymnus* sp. and *Brumus suturalis* F. are predaceous on *Oligonychus coffeae*. Other potential predatory coccinellids recorded from India are *Menochilus sexmaculatus*, *S. pauperculus*, *Coccinella septumpunctata*, *Chilochorus nigratus*, *Brumus suturalis* etc. Seven species of *Oligota* are also reported to feed on different species of mites. A female lays upto 300 eggs and each grub and adult consumes 10-20 mites per day.

Chrysopids are another group of insects, which feeds on mites. *Chrysoparla carnea* is reported to consume 1000 to 1500 citrus red mites daily but fail to complete its life cycle on mite diet. *Chrysopa vulgaris* have known to have better searching ability than *Stethorus* and consume 30-50 European red mite larvae per hour. The insects belonging to families Coniopterygidae, Anthocoridae and Miridae actively predates upon tetranychids. Among the former, *Conioptyx vicina*, *C. vicina*, *Conwentizia hageni* and *C. psocifomes* are reported to feed on citrus mite, six spotted mite and European red mite in orchards. *Anthocous muscilus*, *A. neuromous* and *Orius* sp. are known predators of *P. ulmi*, *T. urticae* and *P. citri*, respectively. *Blephanidopters angulatus* is also one of the major predators of mites in orchards. Nabidae and Lygaeidae bugs are also reported in literature, as mite feeding insects but their economic importance has not been calculated. Few species of thrips, *Scolothrips sexmaculatus*, *S. indicus*, *S. longicornis* have been known predators of tetranychids which reduce the pest population rapidly. The larva of *Haplothrips faurii* consumes approximately 143 eggs of European red mite with in 8-10 days of its development. A larva of ceicidomyed fly, *Arthroconodox occidentalis* consume 380 mites in 17 days. Flies of other dipteran families like Dolichopodidae, Empididae and Syrphidae are common feeders of tetranychid mites. Spiders are another important group of mite predators. More than 30 species are known to predate upon mites but cultures of these spiders are neither feasible nor economical. Among these, Araneidae (*Gasteracantha* sp.), Lycosidae (*Pardosa* sp.), Salticidae (*Salticus* sp.), Oxyopidae (*Oxyopes* sp.), Tetragnathidae (*Tetragnatha* sp.) and Clubionidae (*Clubiona* sp.) are reported as efficient predators of phytophagous mites.

2.6.6 Botanical Pesticides

Solutions containing petroleum-based horticultural oils, vegetable oils or agricultural soaps can be applied to control mites.

Table 2.7: Insect Predators Associated with Mites

Insect	*Mite*	*Insect*	*Mite*
Stethorus sp.	*T.urticae, S. cajani*	*S. pauperculus*	*T. macfarlanei, T. papayae, O. indicus, E. orientalis*
S. gilvifrons	*O. indicus*	*S. punctum*	*Tetranychus* sp.
S. keralicus	*Raoiella* sp.	*S. tetranychi*	*R. indica, T. urticae*
S. parcempunctatus	*R. indica*	*S. punctillum*	*T. urticae*
Oligota sp.	*Tetranychus* sp.	*O. oviformes*	*T. urticae*
Scymnus gracilis	*O. indicus, T. urticae*	*Coccinella* sp.	*O. indicus*
Coccinella undecimpunctata	*Tetranychus* sp.	*C. septempunctata*	*P. latens*
Menochilus sp	*Tetranychus* sp.	*Chilochorus sp*	*Tetranychus* sp.
Brumus suturalis	*O. indicus*	*Micromus timidus*	*T. urticae*
Clanis sorews	*Tetranychus* sp.	Spiders	*Tetranychus* sp.
Adalia sp.	*Tetranychus* sp.	*Jauravia* sp.	*Tetranychus* sp., *R. indica*
Chrysoparla carnea	*E. orientalis*	*Chrysopa vulgaris*	*Panonychus ulmi*
Scolothrips sexmaculatus	*T. urticae, P. ulmi*	*S. longicornis*	*T. urticae*
S. indicus	*T. macfarlanei, T. neocaledonicus, O. indicus, E. orientalis*	*Cryptothrips nigrripes*	Hibernating mites
Haplothrips faurii	*P. ulmi*	*Anthocoris sp.*	*P. ulmi, T. urticae*
Arthroconodax occidentalis	Six spotted mite		

Botanical pesticides could increase the effectiveness of natural predation as they are harmless to natural enemies. Spider mites and eggs are killed by suffocation when the oil or soap solution smothers them. Azadiracditin and NSKE have been shown to possess antifeedant and growth inhibiting effects on large number of mite species. Neem oil (3 per cent), NSKE (5 per cent) are found effective against spider mites. Garlic oil, neem oil and NSKE were also found promising against some eriophyid mites especially coconut mite. Extreme care should be taken with the use of these types of products to limit the chances for phytotoxicity. Petroleum spray oils (PSOs) of various qualities are used to suffocate mite pests especially *Panonychus ulmi* and *Tetranychus urticae* in horticultural crops (Davidson *et al.*, 1991; Herron *et al.*, 1998). Low phytotoxicity, narrow range, no resistance, negligible health and environmental concerns have contributed to increase in their use.

2.6.7 Antimetabolites

A number of new miticidal compounds have been developed within the past fifteen years. These include avermectins, pyroles and milbemycins. Pest mortality results from disruption of metabolism within nerve cells of pest mites. Abamectin, an avermectin, is a mycelial extract of *Streptomyces avermitilis*. Chlorfenapyr is a synthetic pyrole that has proven extremely effective at suppressing populations of spider mites. Unfortunately, chlorfenapyr exhibits avian toxicity so ways should be found out to minimize birds' exposure to its residues. Milbemycins are closely related to the avermectins. The milbemycins can be produced by a variety of *Streptomyces* and originally differed from the avermectins only in the C-13 position.

The avermectins are reported to act by blocking neuromuscular transmission. This blockage results in immobilization of the mites. Several groups of investigators have indicated gamma-amino-butyric acid (GABA) release at the target system. GABA is an inhibitory neurotransmitter in vertebrates as well as invertebrates. It has been hypothesized that the avermectins cause a specific prolonged release of GABA.

Avermectin was evaluated against Tydeid mites (*Orthotydeus californicus, O. caudatus*), which are regarded as a quarantine pest by regulatory authorities on persimmon fruit in New Zealand. The most effective treatment is early season applications of mineral oil or avermectin against mites. Other mite species against which avermectin showed its potential are:

Tetranychidae

Tetranychus urticae, T pacificus, T macfarlanei, T turkestani, Panonychus ulmi, P. citri, Oligonychus pratensis, O. punicae, Eutetranychus hicoriae, Byrobia praetiosa.

Eriophyidae

Phyllocoptruta oleivora, Eriophyes sheldoni, E. erinea, Epitrimerus pyri, Aculopes lycopersici.

Tenuipalpidae

Brevipalpus lewisi, B. phoenicis, Dolichotetranychus floridanus.

Tarsonemidae

Steneotarsonemus bancrofti, S. ananas, S. pallidus, Acarapis woodi.

A fatty acid or mixture of fatty acids can be combined with one or more avermectins or related compounds to achieve synergistic control of pests.

2.6.8 Fungal Pathogens

Fungal pathogens of eriophyids and spider mites are known which often play an important role in the regulation of natural mite populations and are sometimes able to decimate populations of phytophagous mites (Vorce, 1881). Studies are being conducted to develop some of these fungi as commercial acaricides.

2.6.8.1 Entomophthorales

These are host specific usually one or a few very closely related hosts can be infected. Non-target species particularly phytoseiid mites are not susceptible to fungal infection. They are obligatory pathogens, which means that they can only grow on living hosts. Insects and mites infected by these fungi are usually not killed before all available nutrients have been utilised, in contrast to hosts killed by Deuteromycetes where mycotoxins often cause host death prior to complete colonisation of the body.

Initially, *Neozygites* species was observed to cause adult mortality (32-95 per cent) in the citrus red mite, *Panonychus citri*. Subsequently, several other species of spider mites like *Eutetranychus banksi*,

Tetranychus urticae were found susceptible to *Neozygites* sp. Among mite stages, immature stages of *T. urticae* are more susceptible to *N. floridana* than adult mites and females. Among adults, females are more susceptible to infection than males. Infection is due to direct penetration of the fungus through the host cuticle by a combination of enzymatic and mechanical processes.

A major drawback in evaluating the potential of these fungi as biocontrol agents is that it can be cultured only inside the arthropod hosts. Dissemination of the fungus as infected mite mummies is a possible way to spread the pathogen in a spider mite population. This method has been patented for the biological control of *T. urticae* in field crops and greenhouse crops. In this method, cultures of *T. urticae* are infected on Lima beans with the fungus by adding field-collected, infected mites. Dead, infected mites are then collected and stored over calcium carbonate at 4 °C or in a freezer until they are needed for application. However, large-scale production of mummies of the mite is more problematic.

2.6.8.2 Deuteromycetes

In this group, *Hirsutella* sp. infections are known from eriophyids. The citrus rust mite, *Phyllocoptruta oleivora* was found most susceptible on grapefruit as from 5,000 mites on a single grapefruit it showed sudden decimation in its populations (reduced to almost zero) due to *Hirsutella thompsonii*. The spores possess a mucous coat that facilitates adhesion to the host cuticle (Baker and Nemizig, 1968). The conidia germinate and penetrate all body parts in eriophids, although on spider mites, penetration is usually through the legs. Over 50 species of entomogenous fungi have been reported in the genus *Hirsutella*, but only few have been reported as pathogens of eriophyids. *H. thompsonii* and *H. necatrix* are two isolates which are effective against spider mites but differ in their pathogenicity. One of the proteins, Hirsutellin A (HtA), inhibits cell growth causing cells to become hypotrophied and disrupting internal organelles and membranes. It caused 100 per cent mortality in *P. oleivora* at a concentration of 100 mg/ml.

H. thompsonii can easily be cultivated on variety of agar-based and liquid media, in contrast to most Entomophthorales. The fungus was highly pathogenic to *T. cinnabarinus* when it was grown on potato-dextrose agar or on sterile wheat bran, and to *Eutretranychus orientalis* when grown on potato-dextrose agar..

H. kirchneri was infective towards the eriophyoid *P. oleivora, E. orientalis, Panonychus citri, T. cinnabarinus* (Tetranychidae), and to some degree towards *Hemisarcoptes coccophagus* (Hemisarcoptidae). No infectivity was found towards *Polyphagotarsonemus latus* (Tarsonemidae), *Rhizoglyphus robini, Tyrophagus putrescentiae* (Acaridae) and *Typhlodromus athiasae* (Phytoseiidae). The fungus has also no effect on beneficial insects such as the coccinellids, *Coccidophilus citricola* and *Lindorus lophantae*

Paecilomyces eriophytis on the black currant mite *Cecidophyopsis ribis* and *Cephalosporium* sp. on *Phytoptus avellanae* (Eriophyoidea) was also found effective. The latter's isolate *C. ribis* also showed promising result against *P. ulmi, Aceria hippocastani, Cecidophyes galii, A. vaccinii* and *Polyphagotarsonemus latus*.

Fungal pathogens seem promising against mites as they thrive under warm and humid conditions. The pathogenicity of conidia of *Beauveria bassiana, H. thompsonii* and *Paecilomyces fumosoroseus* were tested in the laboratory under controlled temperature and humidity conditions and in the greenhouse. All fungus species tested were able to infect the mites; higher doses of conidia resulted' in a faster death of the mites, while density of the mites on the leaves also affected disease incidence. For example, mortality caused by *B. bassiana* was fastest among densities between 65 and 125 mites per (bean) leaf. The broad mite has also been found in association with *H. nodulosa*.

One of the earliest experiments in which a fungus was tested against a Phytophagous mite was probably a field application of *B. bassiana* spores (0.5 per cent) for the control of *T. urticae* which caused 71 per cent mortality. Among 152 different isolates of the fungi *B. bassiana, B. brongniartii, Beauveria* sp., *Metarhizium* sp., *Paecilomyces lilacnus* and *P. farinosus* tested, only the *Beauveria* spp. tested caused mortality between 35 and 95 per cent.

2.6.9 Viruses

Virus diseases are known in only a few mites, namely, the citrus red mite and the European red mite (Steinhaus, 1959; Beavers and Reed, 1972; Reed, 1981). In both cases, non-occluded viruses play an important role in the regulation of mite populations in citrus and peach orchards, respectively, but application of these viruses as biological control agents does not seem feasible.

2.6.10 Bacteria

Few bacteria have been reported as pathogens of the Acari but in recent years research has been concentrated on intracellular organisms such as Wolbachia that may cause distored sex ratios in offspring and incompatibility between populations. The effect of *Bacillus thuringiensis* on mites is also reported in literature. Field applications were successful against the *P. citri, T. pacificus* and *T. urticae* (Poinar and Poinar, 1998). Young instars are especially susceptible, since these have a high growth rate. Physiological processes that occur in young organisms require rates of RNA synthesis than in older, slower growing adults. Thuringiensin is toxic for a wide range of organisms, including spider mites.

2.6.11 Chemical Control

The acaricide should possess the properties desired of mite control agents with respect to environment compatibility including minimal side effects on nontarget species. The type of action of an acaricide often limits its use as an acaricide (Hoyt *et al.*, 1978; Hoy *et al.*, 1979). A toxicant that kills mites only by direct contact during application is relatively ineffective for mite control on dense or low growing crops. Coverage of all aerial parts of the plants is normally necessary on other crops. In contrast, acaricides that penetrate through leaves or those that are toxic in the vapour phase require a minimum of spray coverage to achieve adequate mite control.

First treatment should be applied to all host plants in an area when populations are low. If the first treatment kills only the active stages, then for effective control, the second treatment must be applied after all the eggs are hatched and before developing mites begin to deposit eggs (Dhooria, 2003).

Sulphur is an effective acaricide, which is used, in different forms against different mite pests. It is not toxic to the applicator or to those who consume crops on which it is applied. It is not hazardous to predatory mites. In addition to sulphur, various synthetic organic compounds differing markedly in their chemical configurations possess acaricidal compounds. These include organophosphates (TEPP, parathion, phorate, carbophenothion, chlorpyriphos, azinphosmethyl), carbamates (Formetanate, carbofuran, aldicarb), chlorinated hydrocarbons (endosulfan, pentae, dicofol), Nitrophenol derivatives (Dinocap, binapacryl, dinobution), Pyrethroids (Fenpropathrin), Sulfonates (Ovea, fenson), Sulfones (Tetradifon), Sulfides (Chlorbenzide), sulfites (Aramite, propargite (Omite), organotins (plictran), formamdines (chlordimefom), Benzoxyhydroxymites (Benzomate) and Quinozalines (Oxythioquivix). In addition, certain novel acaricides have come up which are effective against phytophagous mites. These are Abamection (discussed under separate heading), acrinathrin (pyrethroid), amitraz, azocyclotin, benzoximate, crofentizine, cyhexation, diafenthiuron, dienochlor, ethion fenazoquin, fenbutatin oxide, fenothiocarb, flucyloxuron (benzoyl urea), formetanate (carbamate) milbectin, pyrimidifen and tebufenpyrad. Different acaricides/ chemicals tested against various pest species are provided in Table 2.8.

2.7 Bioecology of Important Phytophagous Mites

Phytophagous mites are discussed under different agricultural crops like cereals, feed and fodder, fruits, millets, ornamentals, pulses, vegetables and other economically important crops.

2.7.1 Vegetables

2.7.1.1 Brinjal

***Tetranychus telarius* Linn.**

The two spotted spider mite is polyphagous in nature with wide distribution in India. Female is generally scarlet red in colour, oval and pyriform. Eggs are laid on the underside of leaves. About 40-60 eggs are laid per female. Incubation, larval, and protonymphal period lasts for 5-8, 3-4 and 3-5 days, respectively. Deutonymphal period varies between 3-5 days for females and 1-4 days for male. Male lives for 3-9 days while female lives for 8-14 days. An adult female aestivates in summer and comes out with rains in Rajasthan. Spherical eggs are laid parthenogenetically and develop into males. Second generation is sexual. Fifteen generations are completed from July to March.

Young and adult mites feed on undersurface of leaves by sucking the cell sap resulting into discoloration of leaves. Webbing on the leaf surface interfers with the plant growth and gradually leaves dry up. Fruit setting is reduced.

Table 2.8: Different Acaricides/ Chemicals Tested against Phytophagous Mites

Sl.No.	Mite Species	Crop	Acaricides
1.	*Tetranychus urticae*	Okra	Wettable sulphur (0.5%), Dicofol (0.04%), Dioxathion (0.1%), Replin (1.5%), malathion (400 ml/acre)
		Brinjal	Sulphur (0.2%), Dicofol (0.09%), tetradifon (0.03%), endosulfan (0.075%), monocrotophos (0.04%), malathion (400 ml/acre)
		Beans	Dicofol (0.04%), ethion (0.1%), Neemark (0.5%), Replin (1.5%)
		Cotton	Dicofol (0.03%), monocrotophos (0.04%)
		Cucumber	Malathion (250 ml/acre)
		Apple	Tebufenpyrad (25 ppm)
		Rose	Flufonoziron (1.5ml /l)
		Marigold	Benomyi (0.25 Lb)
		Jasmine	Dicofol (2 ml/l), wettable sulphur (3g/l)
2.	*T. cucurbitae*	Brinjal	Plictran (0.03%), wettable sulphur (0.15%)
3.	*T. equitorius*	Mulberry	Phosalone (0.1%), endosulfan
4.	*T. macfarlanei*	Brinjal	Sevisulf (Carbaryl+ sulphur) (0.1%)
		Okra	Dicofol (0.04%), monocrotophos (0.04%)
		Cotton	Dicofol (0.03%), monocrotophos (0.04%)
5.	*T. neocalendonicus*	Brinjal	Plictran (0.03%), sulphur (0.15%), dicofol (0.04%), tetradifon (0.03%), endosulfan (0.075%)
		Rose	Dimethoate (0.1%)
6.	*T. ludeni*	Cowpea	Dicofol (0.05%), ethion (0.1%)
7.	*T. telarius*	Brinjal	Dimethoate (500ml), methyl oxydemeton
8.	*T. turkestani*	Cotton	Propargite (1500 g a.i./ ha)
9.	*Oligonychus indicus*	Sorghum	Ethion, Dicofol, Tetradifon (0.05%)
		Sugarcane	Malathion (0.075%), Dicofol (0.05%), monocrotophos (0.05%), fenthion (0.075%), methyl oxydemeton, dimethoate
10.	*O. mangiferus*	Mango	Dicofol (0.04%), amitraz (0.04%), sulphur (0.25%), Dinocap

Contd...

Table 2.8–Contd...

Sl.No.	Mite Species	Crop	Acaricides
11.	*O. sacchari*	Sugarcane	Dicofol (0.08%), malathion (0.05%), diazinon, phorate, monocrotophos (0.075%)
12.	*Eotetranychus hicoriae*	Guava	Sulphur (0.2 kg a.i./ha), Dicofol (0.3 kg a.i./ha)
13.	*Eutetranychus orientalis*	Citrus	Soil application of aldicarb (0.5g/ nursery plant), disulfoton (2g/ plant), fenthion, monocrotophos, methyl oxydemeton, permethrin
14.	*Panonychus ulmi*	Apple	Dinocap, Dicofol (0.05%), monocrotophos, fluvalinate (0.025%)
15.	*Petrobia latens*	Barley	Ethion (0.05%)
		Wheat	Ethion (0.05%), Dicofol (0.04%), dimethoate (0.15%), methyl oxydemeton (0.1Kg /ha)
16.	*Paratetranychus citri*	Citrus	Wettable sulphur (1%), dimethoate (0.03%), Dicofol (0.05%), malathion (0.05%), lime-sulphur-wash (1:2:16)
17.	*Polyphagotarsonemus latus*	Chilli	Triazophos, phosalone (0.1%), propargite (2l/ha), ethion (0.05%), wettable sulphur (0.5%), fenpropathrin (0.01%)
		Capsicum	Dicofol (0.025%)
		Citrus	Propargite, sulphur (15-100 kg/ha), carbophenothion, avermectin
		Tea	Dicofol
		Mungbean	Dimethoate (0.05%)
18.	*Brevipalpus californicus*	Guava	Carbophenothion (0.15%)
		Sweet lime	Malathion (0.05%)
19.	*Brevipalpus sp.*	Citrus, grapevine	Dicofol (0.03%, 0.05%), cypermethrin (0.005%)
20.	*B. phoenicis*	Guava	Dicofol (0.05%), cypermethrin (0.005%)
21.	*Raoiella indica*	Coconut	Monocrotophos (0.04%), dimethoate (0.03%), phosphamidon (0.05%), Dicofol (0.03%), tetradifon (0.03%)

Contd...

Table 2.8–Contd...

Sl.No.	Mite Species	Crop	Acaricides
22.	*Aceria guerreronis*	Coconut	Soil treatment of monocrotophos (5-10 ml/plant), foliar spray of triazophos, carbosulfan, monocrotophos (0.03%), Dicofol (6 ml/ l water), dimethoate (0.03%), wettable sulphur (0.2%), methyl demeton (4 ml/l)
23.	*A. jasmini*	Jasmine	Dicofol (0.04%)
24.	*A. litchi*	Litchi	Dimethoate, wettable sulphur
25.	*A. cajani*	Pigeonpea	Dicofol (0.04%), methyl-o demeton (0.05%), monocrotophos (0.05%), Ethion (0.1%), NSKE (5%), neem oil emulsion
26.	*A. mangiferae*	Mango	Dimethoate/ formothion (0.2%)
27.	*Eriophyes sheldoni*	Citrus	Lime sulphur (2%), triazophos
28.	*Phyllocoptruta oleivora*	Citrus	Dicofol, monocrotophos, amitraz

Tetranychus cinnabarinus (Boisduval)

It is a synonym of *T. urticae.* It is also called two-spotted spider mite. In India, it is found in all the States and is very common in Bihar, Mysore, Rajasthan, Uttar Pradesh, Punjab and Haryana. It is a polyphagous pest and is known to feed on 183 species of plants including a number of vegetables. Both nymphs and adults cause damage by sucking the cell sap from undersurface of leaves. It results discoloration of infested leaves, which turn bronze and dry up. Severely infested plants remain stunted with reduced flowering and fruiting. A copious web is formed on the leaves giving an unhealthy appearance. The mite is active from March-October and passes winter as a gravid female. When the season warms up in March, it spins webs on the undersurface of leaves. It lays 60-80 eggs per female and oviposition lasts for 10.87 days. Eggs are spherical and hatch in 2-6 days. The young larvae and nymphs are light brown in colour. The duration of larval, protonymphal, deutonymphal period is 1.89, 1.89, and 1.83 days, respectively. Development from egg to adult is completed in 8.5 days. The body of female is oval, pyriform and is variable in colour. It may be red, greenish amber or rushy green. Males and females live for 9.85 and 12.95 days, respectively (Khanna, 1991). It completes 32 generations in a year.

Tetranychus neocaledonicus **Andre**

It is commonly known as red vegetable mite. Body is carmine or reddish in colour. Nymphs are green with dark specks on body. It is usually found on the undersurface of the leaves. Mite feeding leads to chlorophyll damage resulting in pale yellowish appearance of plant. It also attacks okra, sponge gourd, bitter gourd, beans, tomato, potato and rose but infestation is of lower magnitude.

Oligonychus coffeae **(Nietner)**

The attacked leaves show discoloration and curling of leaves. The larvae and adults live under a web. In North India, it remains inactive till February, after which it copulates and lay eggs. A female lays 80-90 eggs in its lifetime, which hatch in 4-5 days. One life cycle is completed in 8-9 days. High multiplication rate is observed during April-June. *Clanis sorews.*, *Scymnus* sp. and *Brumus suturalis* F. are predaceous on this mite.

Aceria lycopersici **(Wolff)**

It is commonly known as brinjal/tomato erineum mite. It is a minor pest of solanaceae family, which causes erineum (excessive growth of hairs) on leaves. Infestation is throughout the year and generally occurs on undersurface of leaves. In addition to brinjal, its severe infestation may cause damage to tomato, chillies and potatoes and can easily kill the plants.

2.7.1.2 Capsicum

Tetranychus cinnabarinus **(Boisduval)**

On capsicum a female lays 7.5-eggs/ day with 70.3 eggs laid during its lifetime. The larva, pronymph and deutonymph takes 2.55, 2.34 and 2 days, respectively, to moult into next stage. The preoviposition, oviposition and post oviposition period is 2.24, 8.18 and 2.63 days, respectively. Longevity of the mite is 13.7 days. On an average, 11.56 days are needed to complete its development from egg to adult.

2.7.1.3 Chillies

Polyphagotarsonemus latus **(Banks)**

It is commonly known as leaf curl mite/ yellow mite/ broad mite (Karuppuchamy and Mohansundaram, 1987). It is reported to complete its development in 4.57 days at 27°C and in 5.71 days at

22.5°C. A female lays an average of 5-8 eggs. It is associated with many plant-feeding insects like aphids and white flies. These mites can attach themselves to the legs of white flies and use them as a carrier for their dispersal (Natarajan, 1988; Flechtmann *et al.*, 1990).

2.7.1.4 Cucurbits

***Tetranychus cucurbitae* Rahm. and Sapru**

The mites live beneath the thick web on the leaf surface. Parthenogenesis is common and the resulting progeny consists of males only. A female lays 61-93 eggs randomly on the web on lower surface of leaves. Incubation period varies from 3-30 days depending upon climatic conditions. The larval period is from 2-20 days for males while 2-27 days for females. Mites damage the plants by desapping and by covering the leaves with thick web on which soil particles get adhered. Photosynthetic activity of such leaves gets highly reduced. It adversely affects flowering and fruiting (Yadav *et al.*, 1987).

2.7.1.5 French Beans

***Tetranychus neocaledonicus* Andre**

It is a commonly occurring web spinning mite on French beans. The total developmental period for males and females is 10.44 and 10.19 days, respectively.

***Tetranychus ludeni* Zacher**

It also known as bean spider mite or red legged spider mite or carmine mite. The early infestation of mite causes yellowing of leaves followed by necrotic patches. During heavy infestation, bronzing effects on leaves are very clear. This mite also occurs in mixed population with *T. cinnabarinus*. This polyphagous mite lays 165 eggs/ female on French beans with male: female ratio of 1:27. Pre oviposition and developmental period is recorded as 1.54 and 11.5 days, respectively.

2.7.1.6 Okra

***Tetranychus cinnabarinus* (Boisduval)**

It is most commonly occurring spider mite (now known as *T. urticae*) in India causing economic losses in vegetable crops. Life cycle is completed with in 9-15 days, depending upon the host and temperature (22-28°C). Temperature with in the range of 16-20°C prolongs the duration of generation to 29 days. Eggs are laid along the veins and midrib of the leaves. Eggs are soft, smooth and spherical in shape. Female lays 8-12 eggs in a day. In her life, a female lays 75-120 eggs. Egg hatching period is 4-5 days whereas larval, protonymphal and deutonymphal period lasts for 1-2 days each.

***Tetranychus neocaledonicus* Andre**

It is commonly known as red vegetable mite whose nymphs are green with dark specks. This tetranychid mite is common in vegetable crops, which is considered as serious pest along with *T. cinnabarinus* in many parts of the country. Preoviposition period is recorded as 12h. Larval and protonymphal period is 1 and 1.5 days and life cycle is completed in 4.5 to 8 days.

***Tetranychus macfarlanei* Baker and Pritchard**

This mite is a serious pest of vegetables in Northern India. The infestation is more on mature crops. It builds up small colonies on the under surface of leaves. The sucking of plant cell sap by mites cause yellowing of leaves followed by drying.

2.7.1.7 Tomato

***Aceria lycopersici* (Wolff)**

This mite has worldwide distribution and commonly referred to as tomato/brinjal erineum mite. The first sign of damage to tomatoes are the curling of the lowest leaves and appearance of a silver gloss on the lower surface. Later these leaves become bronze coloured, withered and eventually die. They also feed on the stem, which results in loss of surface hairs on stems. The colour of the stem changes from green to brown and cracks appear on the surface.

2.7.1.8 Sem

***Tetranychus cinnabarinus* (Boisduval)**

As with other vegetable crops, *T. cinnabarinus* is the most frequently occurring mite in sem. Life cycle is completed in 12 days and females lived significantly for a longer time than males. Fotedar (1981) gave the technique for culturing this mite on sem seedlings and studied the combinations of fertilizers for its optimum development.

2.7.2 Cereals and Millets

2.7.2.1 Rice

***Steneotarsonemus spinki* Smiley**

It is a tarsonemid mite, causing severe damage to rice crop in some parts of the world. It is most abundant during heading to milking stage in the field. Mite infestation leads to increase in percentage of empty grains that ultimately leads to grain loss. This mite carries spores of rice sheath rot fungus, which causes brownish spots on rice sheath and grains; damage termed as 'Sterile grain syndrome'. It completes its life cycle in 17-18 days at 25°C but at 30°C, it takes 3-4 days to complete its life cycle. However, at 30°C, longevity decreases drastically to 5 days as compared to 15 days at 25°C. A female lays 30-70 eggs in her lifetime. This mite is highly sensitive to relative humidity. When relative humidity reaches below 40 per cent, all the mites die with in 4-5 hours.

2.7.2.2 Sorghum

***Oligonychus indicus* Hirst**

It is a tetranychid mite which is greyish green in colour. It often causes serious damage in South India, Haryana and Punjab. Injured leaf shows red spots, which become larger as mite feeding increases, ultimately making the whole leaf reddish in colour. Such leaves wither and fall prematurely. The mites live in colonies under a delicate web on the lower surface of leaves. A female lays 35-69 eggs in its lifetime with 1-7 eggs/day. Mites are seen on both surfaces of leaves but abundant on lower side. The incubation period lasts for 9-29 days depending upon climatic conditions. The nymphal period is 1½–19 days for males and 3-24 days for females. Thirty generations are completed in one year. Parthenogenesis is common mode of reproduction. The oviposition period lasts for 10-15 days during April to September and 17-39 days during October to December.

The lady bird, *Scymnus gracilis* Motsch; the anthocorid, *Triphleps tantilus* Motsch, the lygaeid, *Geocoris* sp. and a gamasid mite feed on the nymphs and adults of *O. indicus*. *Scolothrips sexmaculatus* Perg. and grubs of staphylinid, *Lipophaena gracilipes* destroy its eggs.

2.7.2.3 Wheat

***Petrobia latens* (Mulker)**

It is commonly known as brown wheat mite. It is dark brown/ greenish brown in colour with pale yellow legs and a major pest of wheat. Adult mite and its developmental stages are found on the leaves and spikes of wheat. During heavy infestation, leaves dry up from tip downwards, start showing yellowing appearance and ultimately dry out. Heavily infested plant gives sickly appearance. There are five developmental stages from egg to adult. Each stage except egg is followed by a quiscent stage. The mite moulted at the termination of each quiescent period. The average pre-oviposition period is 1.8 days while one generation is completed in about 20 days. A female lays either non-diapausing winter eggs (8-13 /day) or diapausing summer eggs (2-5/ day) at one time with a total of 29-110 winter or 11-43 summer eggs per female. The winter eggs are deposited mainly on dry leaves, which are brick red in colour. Eggs are laid singly or in batches. While the summer eggs are white and laid on soil clumps in clusters. These diapausing eggs remain dormant throughout summer and rainy season (Sharma and Srinivasa, 2004). The eggs hatch in 8-13 days. The larva has 3 pair of legs and larval period is 1-2 days followed by a quiescent stage. The protonymphal and deutonymphal period is 0-3 and 1-4 days, respectively, and possesses four pair of legs. The longevity of adult female is 20 days. Soil application of zinc and boron reduced the mite population.

2.7.3 Pulses

2.7.3.1 Mungbean

***Polyphagotarsonemus latus* (Banks)**

It is commonly known as yellow mite or broad mite. It is a pest of several agricultural crops throughout India. It infests apical 2-3 leaves and buds. Both nymphs and adults suck the cell sap from the leaves. In heavy infestation, there is downward curling of leaves; leaves become dark reddish brown with stunted growth of shoots. Female laid oval, white eggs on the undersurface of leaves. Eggs are studded with whitish tubercles. Eggs hatch in 1-3 days. Young larvae have six pair of legs. Fully fed larva enters the quiescent stage in 3-

11 days to change into an adult. Adult is 1.5 mm long, yellow in colour with white stripes dorsally.

2.7.3.2 Pigeon Pea

Aceria cajani **Channa Basavanna**

This species is worm like with two pairs of legs and acts as vector for pigeonpea sterility mosaic disease (Seth, 1962; ChannaBasavanna, 1966). Although a single mite is enough to transmit this disease (Janarthan *et al.*, 1972), presence of 20 mites provides 100 per cent disease transmission. The mite infests the ventral surface of tender leaves. Three types of symptoms (Nene, 1972; Reddy *et al.*, 1998) are generally seen:

1. *Complete sterility*: Early infestation (before 45 days) leads to severe mosaic on leaflets and complete suppression of flowering and fruiting. Diseased plants show dwarfing of leaves, bushy habitat and yellowish green leaves.
2. *Partial sterility*: In late infestation (beyond 45 days), few branches show mild mosaic on leaflets and such branches do not bear flowers and fruits.
3. *Ring spots*: Leaves are characterized by green islands surrounded by a Chlorotic halo, which disappear as the plants mature, and with no significant effect on flowering.

Oldfield *et al.* (1981) studied the biology of this mite. It has short life cycle (< two weeks) that includes egg, two nymphal and adult stages. Eggs are milky white, normally found attached to filamentous trichomes at the growing tips of the plants. Female lay 1-3 eggs daily which takes 3-4 days to hatch.

2.7.4 Horticultural crops

2.7.4.1 Fig

Eotetrancyhus hirsti **Pritchard and Baker**

It is commonly known as fig spider mite. This mite is host specific. It is greenish yellow mite with black blotches all over dorsum. Transparent green patches are produced on the under surface of leaves which turn yellowish green than brown with rough and dry texture. Such leaves alongwith fruits drop prematurely, often resulting in complete defoliation of plants.

2.7.4.2 Mango

Aceria mangiferae Sayed

The bud mite sucks the sap from the buds and cause necrosis of the tender tissues. When poulation is high, the bud may be killed. Highest population is found during February and then in July. In between population remains low. In August-September, there is a further decrease in population, which remains low during October-December. A rapid increase in January results in the peak population in February. Remove and destroy all the panicles bearing infested inflorescence to check its population in field.

Oligonychus mangiferus (Rahman and Sapru)

It is commonly known as Mango red spider mite. The mite is dorsolaterally red and mediodorsally pale red in colour. Sucking of leaf sap results in the appearance of characteristics yellowish spots. The middle area of lamina becomes much yellowish with less browning.

2.7.4.3 Litchi

Aceria litchi (Keifer)

The leaf curl mite has been reported from the entire litchi growing regions and it is commonly known as litchi erineum mite. It is reported from India (Singh, 1988), China, Hawaii and Pakistan. Mites attack newly emerged leaves, young fruits, shoots and flower buds. Affected plant parts on maturity become distorted curl, dry and fall. Small galls or wart like swellings or depressions appeared on the under surface of the infested leaves (Lal and Rahman, 1975). Due to heavy infestation 5-45 per cent leaf damage, 75-100 per cent flower bud damage and 30-50 per cent fruit loss are reported (Singh, 1988).

Life cycle consists of egg, protonymph (2-3 d), deutonymph (5-7d) and adults, which are vermiform, four legged and whitish in appearance. Round, whitish eggs are laid singly on the ventral surface of leaves, which hatch in 2-3 days. Nymphs feed on soft leaves and nymphal stage lasts for 8-12 days. Adult longevity is recorded as 2-3 days. Both nymphs and adults suck the cell sap by lacerating the leaf tissue. Presence of this pest is noticed by the brown velvety growth on the ventral surface of leaves. Infested leaves start curling up, ultimately dry and fall down. Seedlings and young plants are severely damaged. Initially, lower leaves of the tree are infested and gradually reach upper leaves. Both the nymphs and adults have similar appearance except the nymphs are smaller in size. They are whitish in colour with four pair of legs. Adults are 0.15-0.20 mm long.

Three predatory mites and few fungi (Bilgrami, 1982) are associated with this mite, which limits its population. Removal or destruction of litchi leaves either through burning or burying them deep in soil destroys the mite population. Pruning of infested shoots during December is recommended when mites remain in the hibernating stage.

2.7.4.4 Ber

***Larvacarus transitans* (Ewing)**

It is commonly known as ber gall mite as it produces round galls on ber. The distinctive characters are elongated body with three pairs of legs in nymphs and adults. The unfertilized females produce male only whereas fertilized females give normal progeny of males and females. The galls rupture during monsoon and the mites after emerging from galls hide themselves in cracks and crevices of tree trunk and start fresh infestation when new plant growth takes place.

***Eriophyes cernuus* Masses**

It is commonly known as ber woody gall mite. It is a light brownish mite, which is a minor pest of ber. It occurs throughout the year and produce woody galls. The galls are usually on one side of the stem. Ringing the base of developing bud and shoots.

2.7.4.5 Apple

***Panonychus ulmi* (Koch)**

It is one of the important mites associated with apple orchards and has taken a serious pest status in many parts of the world. Mostly it occurs on the upper surface of the foliage and that too mainly on spur foliage. Heavy population leads to reduced fruit size and yield as it interfers with fruit bud formation. Adults, larvae and nymphs suck sap from leaves and discolour them, which results in poor size and colour of the fruits in the current year and reduced fruit set in subsequent years (Hull and Beer, 1989). It overwinters as an egg. Eggs are laid mostly on spurs, small branches, cracks and crevices of the bark. Eggs hatch in early spring. *P. ulmi* completes one generation in 2-4 weeks depending on the temperature. In one year, 4-5 generations are completed by this mite.

Resistance to acaricide reported (Bower, 1977; Chapman *et al.*, 1987; Pree and Wagner, 1987; Welty *et al.*, 1987; Thwatte, 1991; Khajuria, 2000). Population density of this mite can be influenced by manipulation of nitrogen fertilization. Two phytoseiid mites, *Metaseiulus occidentalis* and *Amblyseius fallacis* are reported to feed on this mite. Other mites associated with apple crop are *Tetranychus arboreus* and *Eotetranychus carpini borealis, Aculus schlechtendali.*

Aculus schlechtendali (Nalepa)

It is also known as apple rust mite. It over-winters as an adult in cracks and crevices on the old wood, lateral buds on terminal shoots. Adults emerge in early spring and move to the opening buds. Eggs are laid on the under sides of the leaves which hatch in 10-12 days, depending on temperature. During heavy infestation, 280-290 mites/leaf are recorded. *A. schlechtendali* could be beneficial for the control of tetranychid mites on apple as early season feeding of rust mite make the foliage slightly less susceptible to red mites. Secondly, it provides alternate food supply to predatory mites, which multiply and increase their numbers before *T. arboreus* and other tetranychid mites appear on apple leaves to provide better control.

Tetranychus urticae Koch

Herbert (1981) described its biology on apple. Protonymph and deutonymph completes their development in 2.45, 2.91 days, respectively, at 21 °C. Preoviposition and oviposition period are recorded as 2.2, 26.5 days, respectively.

2.7.4.6 Grapes

Tetranychus pacificus McGregor

It is a tetranychid mite associated with grapes. It mainly occurs along the mid rib and vein of leaves and prefer ventral surface of the leaves. As like other tetranychids, it spins webs and prefers to stay inside the web.

2.7.4.7 Guava

Brevipalpus phoenicis (Geijskes)

It is commonly known as Scarlet mite or reddish black flat mite. It is a dark reddish flat mite with black or green spots and occurs on under surface of leaves. This tenuipalpid mite is a major pest of fruit trees, plantation crops and ornamental plants. Infested leaves turn pale yellow and develop brownish patches. It is commonly found in the field with mixed population *B. californicus* and *B. obovatus.*

2.7.4.8 Pear

Tetranychus urticae **Koch**

This crop has very low tolerance to mite infestation. Few mites per leaf resulted in blackenning of leaves (transpiration injury). Heavy infestation leads to reduced fruit size, fruits per cluster, yield and pre-harvest drop.

2.7.4.9 Mulberry

Tetranychus ludeni **Zacher**

It is commonly known as red legged or caramine mite. Due to continuous feeding on leaves, white specks are formed which gradually enlarge to form large patches. Leaves give rusty appearance and gradually dry. A female lays eggs on the web on the underside of leaves. Freshly laid eggs are brown in colour, which later become whitish. Eggs hatch in about 4 days. Newly emerged larva is creamy white with two prominent red spots on propodosomal region. Nymphal stage lasts for 2 days. It takes 9-10 days from egg to adult stage.

Tetranychus equitorius

Mite is active and cause damage when the temperature is 28-33°C and RH is about 50per cent. Female lays 39-50 eggs during March-June. Incubation period is 5 days while nymphal duration in 4 days. Life cycle is completed in 10 days.

2.7.4.10 Citrus

Eutetranychus orientalis **(Klein)**

It is commonly known as oriental red mite or citrus mite and has a host range of more than 85 plant species including citrus species, peach, pear, amaltas, neem, shisham, papaya, almond etc.

Mites suck the cell sap from upper surface of leaves. Such leaves turn yellow and later drop off. Infestation generally starts along with the midrib and then spreads to lateral veins. The mites also produce webbing on upper surface of leaves. The period from egg to adult stage is 12-13 days (Banu and ChannaBasavanna, 1972). The larval, protonymphal and deutonymphal period is 1.4, 1.5 and 1.5 days, respectively. On an average, a female lays 51 eggs with an oviposition period of 5 days (Kanta, 1971).

Predatory mites, *Amblyseius alstoniae, A. cucumeris, A. hibisci, Pronematus* sp. and *Agistemus* sp. and insects like coccinellid beetle, *Scym-nus gracilis* and thrips, *Scolothrips indicus* feed actively on all the stages of citrus mite.

Eutetranychus banksi (McGregor)

This is a tetranychid mite, recorded from citrus and other fruit crops. The mites live both on the upper and lower surface of leaves. Mite feeding results in white spots on leaves.

Panonychus citri (McGregor)

It is commonly called as citrus red mite. Apart from citrus, this mite is also recorded from rose, almond, pear, peach, cherry and apple. Development from egg to adult is completed in 14-15 days. Female and male longevity varies from 7-11 days and 6-9 days, respectively. Each female lays 50 pearly white, subspherical eggs. It over-winters as an egg in twigs and bark of trees. Two strains of this mite exist in the field. One is diapause strain and other is non-diapause strain. Diapause strain eggs are larger than the eggs of other strain. Non-diapause strain is known to pass the winter in developmental stages. The mite causes stippling, light coloured spots and greyish or silvery appearance on leaves. It feeds both on leaves and fruits causing bronzing or silvering of the surfaces. Mite also suck the sap from green fruits. The mite can be well controlled by the insecticides referred in *E. orientalis.*

Paratetranychus citri Mc Gregor

Mites are considered to be highly destructive pests of citrus plantations. By their constant feeding injury is caused to leaves, tender fruits and green bark. It results in speckled appearance of the leaves. Heavy infestation may result in complete defoliation especially in young nursery plants. Affected fruits become yellow and remain undersized. It was first recorded in South India where it was found along with *Tetranychus sexmaculata* Riley, *T. telarius* Linn. and *T. hindustanicus* Hirst. In Punjab, it is found infesting sweet orange, lemon and grapefruit.

Adult is orange in colour with thick deep brown patches on the dorsal side of its body. Its body is covered with prominent bristles, each borne on a whitish tubercle. Antennae of female are 3 segmented and are bright pink. The newly hatched larva is light yellowish brown and has 3 pair of legs. The protonymph is orange brown and the deutonymph is orange brown with a greenish tinge.

The pest is active throughout the year but it is most active during May-June. A female lays 50 eggs arranged singly along the veins on the underside of leaves. Eggs are round, orange and are embedded in leaf tissues, with ornamentation of threads, which extends from the top outwards. Eggs hatch in about a week. Unfertilized eggs develop into males while fertilized eggs into female and male. The first instar larva has three pair of legs. After 3-4 days, it moults into protonymph with 4 pair of legs. It changes into dentonymph within 3-4 days. After 4-5 days, it transforms itself into an adult mite. Female live for 10 days, life cycle in summer is completed in 17-20 days and the pest passes through several overlapping generations in a year.

***Phyllocoptruta oleivora* (Ashmead)**

It is commonly known as citrus rust mite. It is yellowish in colour with dorsally flattened and fusiform body. It is a serious pest of citrus throughout the world (Rao *et al.*, 1957). High mite incidence results into bronzing of leaves and green growing twigs. Mites attack pea size fruits, such fruits become silvery or reddish brown and bring less price. Fruits also remain small in size. Mite is more active during warm and humid conditions.

A female lays 20-30 eggs in 20 days of its life. Egg are smooth, spherical, semitranslucent and laid in groups. The incubation period and nymphal period is 3 days and 2-11 days, respectively. The life cycle is completed in 10-14 days. *Eriophyes sheldoni* Ewing

It is commonly known as citrus bud mite. It is worm like and pale yellow in colour. It attacks all species of citrus. However, lemon is the most preferred host and is found in all lemon growing areas of the world.

This mite remain hidden under bracts of buds, bud scales, developing flowers and fruits. By feeding of mites in leaf axial buds, the bud scales are blackened and bud gets killed. The growth of twigs and floral parts is reduced. A female lays 50 whitish eggs, which hatch in 3-4 days. Development from egg to adult is completed in 12-33 days.

Brevipalpus phoenicis (Geijskes)

It is commonly called as red and black flat mite. The eggs are elliptical, reddish-orange in colour and are laid singly in sheltered spaces. Its life cycle on citrus is completed in 15-22 days at 23±3°C and 85 per cent relative humidity (Dhooria *et al.*, 2005). Mite infests both leaves and fruits. In case of heavy infestation, diffused chlorotic patches known as 'phoenicis blotch' are observed on leaves and 'ring-disorder' on fruits. Infested fruits looked brown to blackish with irregular corky skin. The mite can be effectively controlled with the usual miticidal compounds used against other mites. The phytoseiid mite, *Typhlodromus homali* Gupta has been reported commonly feeding on this mite.

Brevipalpus californicus (Banks)

It is commonly known as silver flat mite. The body of adult female is oval, red with a dark red pattern along the central line of the body. The eggs are elliptical, bright red and covered with a sticky substance, glued to the leaf surface. Larval and nymphal period is completed in 21-25 days. Female lays one egg per day in her life span of about 25 days (Gupta *et al.*, 1971). The mite has been associated with 'leprosis' type symptoms on citrus leaves and 'nail head rust' on citrus fruits. On fruits, mite feeding results in silvering of the fruit while on twigs and branches, its feeding causes scaly bark–a condition commonly referred to as 'Florida scaly bark'.

Polyphagotarsonemus latus (Banks)

It is commonly known as broad mite or leaf curl mite. The eggs of the mite are oval, elongate and show whitish tubercles on upper surface. The nymphs remain enclosed with in skins of quiescent larvae until the adult is formed. Females and males have longevity of 13 and 12 days, respectively. One generation is completed in 4-5 days in summer and 7-10 days during the winter. Female deposits an average of 3-6 eggs per day. Mites feed on the lower surface of the leaves causing the leaves to become rigid and curled. Feeding injury is confined to young foliage. Mites also attack citrus fruits leading to silvery-white symptoms, which later turn corky (Dhooria *et al.*, 1987; Pena *et al.*, 2002).

2.7.5 Ornamentals

2.7.5.1 Jasmine

Aceria jasmini Channa

It is an eriophyid mite causing economic damage to Jasmine. Mite infestation is high during warmer periods of year (April to September). Colonisation of these mites leads to the formation of erineum (fatty outgrowths) on leaves, resulting in malformation of vegetative and floral parts.

2.7.5.2 Rose

Tetranychus urticae Koch

It is commonly known as red spider mite and associated with yield losses in rose. It prefers under surface of the mature leaves and associated with severe webbing on crop. Incubation and oviposition period varied from 3-4 and 9-10 days, respectively. Development from larva to adult is completed in 4-5 days (Meyer, 1981). Duration from egg to adult takes 11 days at $30 \pm 2°C$ and 28 days at $20 \pm 2°$ C (Baradaran *et al.*, 2001).

2.7.6 Other Crops

2.7.6.1 Jute

Oligonychus coffeae **(Nietner)**

The mite lives under web on the upper surface of leaves. Due to sucking of sap, grown up leaves turn brown, dry and drop.

Hemitarsonemus latus Banks

The yellow mite is widely distributed in all the jute growing tracts. Both adults and young, suck the cell sap from leaves. Leaves remain smaller in size. As a result of attack, leaves get crumpled, curbed and turn coppery brown. Such leaves drop prematurely. The young apical leaves are attacked first and attack gradually spreads to lower leaves, when infestation is high, internodes become shortened.

Eggs are laid on the undersurface of leaves and hatch in about 27-32 hours nymphal stage lasts for 36 hours and quiescent stage for 6-8 hours.

2.7.6.2 Tea

***Oligonychus coffeae* (Neitner)**

This red tea spider mite is one of the most important pests of tea, which is widely distributed in India. Female lays eggs singly on the surface of leaves along the midrib and veins. A female lays about 135 eggs. Eggs hatch in 4-6 days into larvae during April to August. There are two nymphal stages. Both larval and nymphal stage is completed in 6-8 days. Female survives for 29 days while male for 24 days. The mites are elliptical in shape with bright crimson colour on anterior side and dark purple colour on the interior side.

Infestation generally remains confined to the uppersurface of old leaves. In severe cases younger leaves are attacked. Mites make webbing on leaf surface and feed under webs. Infested leaves turn brown and dry. Young plants when infested, may get killed. Mite population remain high from March-June and again from September-October. Several species of coccinellids, chryropids and staphylinids are predaceous on the mite.

***Brevipalpus phoenicis* (Geijskes)**

The adult of scarlet mite is flat, oval and scarlet in colour. Black marks are present dorsally. Parthenogenesis is very common. The bright red eggs are laid on the undersurface of leaves. A female lays about 45 eggs during a span of 40 days. Incubation period is 6-13 days. Life cycle is completed in 21-28 days. Egg hatching, development and total life span period is 9.53, 19.13 and 41.68 days, respectively, on tea leaves. Fecundity and fecundity rate is recorded as 39.71 eggs per mite and 0.95 eggs per mite per day. Infestation is more along the midrib and leaf bases on the lower surface of leaves. The bark and leaf petioles of affected shoots turn brown and dry up.

***Calacarus carinatus* (Green)**

This purple mite is widely distributed in India. The spindle shaped purple body marked with five longitudinal white waxy ridges dorsally can easily identify the mite. Its life cycle is completed in 7-15 days. It infests both sides of leaves and results into coppery discoloration of the leaves. Infested leaves are covered with white cast off skins along with live mites.

Polyphagotarsonemus latus **(Banks)**

It is also known as yellow mite and is distributed throughout the tropics. Mites feed on the lower surface of leaves as a result it gives bronzy appearance. Severely attacked plants remain stunted and gradually die. Infestation is generally restricted to young foliage. Eggs are oval elongate and studded with longitudinal rows of white tubercles. The larvae resemble adults except for smaller size, and absence of pseudo-stigamatic organ and skin annulations. The resulting nymph remains enclosed within the skin of quiescent larvae till adult is formed.

Acaphylla theae **(Watt)**

It is commonly known as tea pink mite. It is a pink colour mite, which causes economic losses in tea. It prefers young leaves and the infested leaves turn pale and curl upwards. Total developmental period is about 7-10 days. Incubation and nymphal period ranges from 2-3 and 4-6 days, respectively.

2.7.6.3 Coconut

Raoiella indica **Hirst**

It is commonly known as coconut red mite. Feeding of the mite causes reddish spots on the leaves. The mite is reddish in colour and colonizes the under surfaces of the leaves in young and adult plants. Young plants of coconut are more severely infested. Mites lay eggs on the underside of coconut leaves in groups. Eggs are oblong, smooth, red and change to larvae in 6 days. Larvae are sluggish and reddish in colour and lasts for 10 days. Nymphal period is recorded as 16 days and one life cycle is completed in 33 days. As a result of attack, leaves turn yellow and plants give sickly appearance.

Aceria guerreronis **Keifer**

It is a serious pest in the entire coconut-growing belt of India. Mites feed on the meristematic region of fruits causing uneven growth of tissues and brown discoloration on skin. There is distortion and stunting of coconut. Longitudinal splits may occur on the outer surface of the husk. Mites also attack flowers, which cause premature fruit drop. The pest is active throughout the year but damage is high during dry periods. The mite is elongated and worm like in appearance with creamy white colour. The body is finely ringed and beset with bristles. Adult measures 250 μ in lengths. Each female lays upto 18 eggs during its lifetime. After hatching, mite passes through two nymphal instars to become adult. It takes 10 days from egg to adult stage. Dispersal to long distances is through wind and within the plantation; it spreads by phoresis and rain.

To minimise the pest inoculum, fallen nuts are destructed by burning or by burying them in the soil. Restrict the movement of mite-infested nuts from place to place to minimize the spread of mites. Spraying all fruit bunches at the crown with neem oil mixed with garlic extracts and soap solution (20 ml neem oil + 20 g garlic emulsion + 5g. soap in 1 lt. water) twice at monthly interval give effective control.

2.7.6.4 Cotton

***Tetranychus urticae* (Koch)**

Apart from vegetable crops, it is an important mite on cotton. Though infestation is not so severe as in vegetable crops, there is no significant difference in completion of developmental periods. Vereau *et al.* (1978) studied the biology of *T. urticae* on cotton. It completes its life cycle in 9-10 days. Freshly laid eggs are shining and light brown in colour. The incubation and larval period is 3.84 and 2.16 days, respectively. The larvae are creamy in colour and possess three pairs of legs. Larva changes into protonymph and deutonymph before moulting into the adult. Female longevity is calculated as 14.7 days, which is similar as found on rose and apple crop. The male is comparatively shorter with pointed abdomen and slender body. Female is larger in size with blunt abdomen.

***Aceria gossypii* (Banks)**

It is also known as woolly mite. The infested cotton leaves become stunted, crinkled, hard and crisp. When infestation is severe, bolls are shed ultimately affecting the crop yield.

***Tetranychus arabicus* Attiah**

In addition to *T. urticae* and *T. cucurbitacearum*, this is another tetranychid mite associated with cotton crop. It mainly occurs on the underside of the leaves but move to the upper surface when population becomes high. Nature of damage symptoms is similar to other tetranychid mites. *Amblyseius gossipi* Elbadry and *Agistemus exsertus* Gonzales are two predatory mites species, which limits its population in cotton crop. Intercropping of onion among cotton decreases the tetranychid population significantly.

2.7.6.5 Mint

***Tetranychus cinnabarinus* (Boisduval)**

T. cinnabarinus belong to tetranychidae family, which has a wide range of hosts. In some parts of the world including India (Punjab) it is given a status of major pest of *Mentha* crop. It is recorded from *Mentha arvensis, M. spicata, M. piperta* and *M. citrata*. Due to mite feeding characteristic blotching of leaves occur due to loss in chlorophyll content. Larval and nymphal period ranged from 4-6 days and life cycle is completed in 10-11 days.

2.7.6.6 Sunflower

***Tetranychus turkestani* Ugarov and Nikolski**

It is also known as strawberry spider mite. 90-142 mites per leaf are recorded during peak season. During high temperature, it is known to develop into high population and interfere with setting and maturing of buds and seed heads.

2.7.6.7 Sugarcane

***Aceria sacchari* Channa.**

It occurs throughout India. It feeds on the inner side of leaf sheath, which induces formation of blisters of varying shape on the inner surface of leaf sheath. Tissues in these blisters are spongy and harbour colonies of mites. All stages of mites can be seen inside them. The blister is light green in the beginning and gradually shows rusty appearance. In later stages of infestation, the outer side of leaf sheath shows swollen scars. The life cycle is completed in 15 days at 21.8-28.9°C.

Oligonychus sacchari

It occurs in Punjab, Uttar Pradesh, Andhra Pradesh, West Bengal and Tamil Nadu. Mites scrap the epidermis and suck the cell sap of leaves. As a result of feeding, white patches are formed on leaves. Colonies, which are parallel to the midrib, are formed, on the underside of leaves. Life cycle is completed in 11-16 days during September-October and 31-48 days in November-February. It is active throughout the year in tropics. Its population remains high during mid-June to October end in the sub-tropics. It is more severe in ratoon crop than on plant crop. Dispersal occurs through wind.

Oligonychus indicus (Hirst)

The sugarcane red leaf mite is most abundant and occurs in Haryana, Punjab, Uttar Pradesh, West Bengal and South India. Adults form fine webs and a number of colonies on leaves. Both nymphs and adult females cause damage by sucking the cell sap from the undersurface of leaves. Red streaks and webbing is observed on the undersurface of leaves. Such leaves turn red and gradually dry up. In severe infestation, 10-15 colonies may be found on a single leaf. Sugarcane varieties with soft leaves are more severely attacked. Damage is more during the premonsoon season.

Female mites spin webs on the underside of leaves in which 35-69 dirty white, spherical eggs are laid in a life span of 12-33 days. Eggs hatch in 3-4 days. Male nymphs gain to maturity through three stages. Their development is completed in 2-3 weeks. The life cycle is completed in 3-6 weeks. High temperature and low humidity are favourable for the development of this pest. For effective management, the alternate host, *baru* grass should be destroyed from and around sugarcane fields.

Schizotetranychus andropogoni (Hirst)

It is commonly called as white patchy or sugarcane leaf spotted mite. The mite has a pale body with dusky spots on undersurface. The mites make small, oval colonies covered with thin webs on the undersurface of leaves. Their number may reach 1300–1500 per leaf. The feeding on leaves causes the appearance of white blotches, which later turn to brown. The affected leaves look sickly and gradually dry up.

2.7.6.8 Bamboo

Schizotetranychus nanjingensis Ma and Yuan

The species constructs nests with a very dense web roofing on the underside of bamboo leaves and feeds on leaf tissue (Yu and Shi, 1991; Zhang *et al.*, 1998). In early stages, the upperside of an infested leaf is full of whitish stippling which becomes yellowish to brown with increase in infestation. A severely infested bamboo forest has a 'burnt' appearance with leaves falling and photosynthesis of the plant greatly reduced.

Table 2.9: Phytophagous Mites Associated with Agricultural Crops

Family	*Mites*	*Host Range*
Tetranychidae		
	Aponychus lalii Prasad	Ber
	A. sulcatus	Narrah^{+}
	Tetranychus arabicus Attiah	Cotton
	T. bioculatus W.M.	Cotton
	T. cinnabarinus (Boisduval)	Cotton^{+}, beans, castor, okra, brinjal, cassava, fenugreek^{+}, marigold^{+}, rose^{++}, spinach^{+}, tomato^{+}, dahlia^{++}, moong^{++}, mash^{++}, moth bean^{+}, lobia^{++}, soyabean^{+}, pigeonpea^{+}, capsicum, mint
	T. cucurbitae Rahm. and Sapru	Beans, brinjal, cucurbits, cabbage, tomato, tamarind, jasmine
	T. cucurbitacearum	Cotton
	T. equitorius	Mulberry
	T. fijiensis Hirst	Coconut
	T. hindustanicus Hirst	Coconut
	T. joellotti Flechtmann	Cassava
	T. laches VUF	Citrus
	T. ludeni Zacher	Castor, beans, brinjal, coconut, potato, pumpkin, mulberry, cowpea
	T. macfarlanei Baker and Pritchard	Okra, cotton
	T. neocaledonicus Andre	Cauliflower^{+}, coconut, fig, French bean^{++}, brinjal, okra
	T. pacificus McGregor	Grapes
	T. sexmaculatua Riley	Citrus
	T. telarius Linn.	Cotton, castor, okra, grapevine, tea, hemp, tapioca, brinjal, citrus, rose
	T. turkestani Ugarov and Nikolski	Strawbewrry, sunflower

Contd...

Table 2.9–Contd...

Family	*Mites*	*Host Range*
	T. urticae Koch	Cabbage+, chrysanthemum+, lettuce+, papaya, pear++, celosia++, okra++, rose+, brinjal+, beans, cotton, apple
	Oligonychus. biharensis Hirst	Areca palm, cassava
	O. coffeae (Nietner)	Castor, jute, okra, tea, coffee, indigo, citrus, brinjal
	O. indicus (Hirst)	Sugarcane++, sorghum+, jowar, maize+, areca palm
	O. mangiferus (Rahman and Sapru)	Jamun, mango+, grapevine++, pomegranate++
	O. oryzae (Hirst)	Rice
	O. pratensis (Banks)	Coconut
	O. punicae Hirst	Pomegranate, grapevine
	O. sacchari	Sugarcane
	Eutetranychus africanus	Citrus
	E. banksi (Mc Gregor)	Ber, citrus, cotton
	E. ceranuus	Katchnar++
	E. citri	Citrus
	E. fremonti Tuttle and Baker	Plum
	E. orientalis (Klein)	Amaltas++, balsum++, castor, chandni++, citrus++, cosmos++, French bean+, kaner++, marigold+, murraya++, papaya+, peach++, pear++
	E. suginamensis	Mulberry
	E. truncatus	Grapevine
	Eotetranychus hirsti Prit Chart and Baker	Fig++
	E. kankitus	Fruit crops
	E. mandensis	Fruit crops
	E. palmalae	Fruit crops
	E. sexmaculatus	Fruit crops
	Bryobia artemisia	Gram+

Contd...

Table 2.9–Contd...

Family	Mites	Host Range
	B. eharai	Chrysanthemum+
	B. neopraetiosa	Barley, wheat
	B. praetiosa	Pear+
	Schizotetranychus andropogoni (Hirst)	Sorghum, sugarcane
	S. baltazarae	Citrus
	S. cajani	Pigeonpea
	S. fluvialis	Pigeonpea
	S. hindustanicus (Hirst)	Citrus
	S. lechrius	Pinkcassia++
	S. nanjingensis Ma and Yuan	Bamboo
	Panonychus citri	Citrus+
	P. ulmi (Koch)	Apple
	Petrobia latens (Muller)	Barley+, citrus, lentil+, wheat+
	Paratetranychus citri McGregor	Citrus, sweet orange, grape fruit
	Amenosius sp.	Coconut
Eriophyidae		
	Aceria cajani Channa.	Pigeonpea+ (Pigeonpea sterlity mosaic virus)
	A crotalariae Channa.	Sunhemp
	A. ficus (Cotte)	Fig+ (Fig mosaic virus)
	A. gossypii (Banks)	Cotton, tea
	A. granati Can. and Massal	Pomegranate++
	A. guerreronis Keifer	Coconut
	A. hibisci (Nalepa)	Okra
	A. jasamini Channa	Jasmine++
	A. litchi (Keifer)	Litchi++
	A. lycopersici (Wolff)	Brinjal+, tomato+
	A. mangiferae Sayed	Mango (Mango malformation disease)
	A. mori	Mulberry
	A. nerii	Kaner+

Contd...

Table 2.9–Contd...

Family	*Mites*	*Host Range*
	A. peucedani (Canestrini)	Carrot
	A. putturudriahi Channa	Cotton
	A. rubifolii	Raspberry[+]
	A. sacchari Channa	Sugarcane (Leaf streak virus)
	A. sheldoni (Ewing)	Citrus
	A. sorghii Channa.	Sorghum, sugarcane
	A. tulipae Keifer	Wheat (Wheat streak mosaic virus), rye, corn
	Abacarus hystrix (Nalepa)	Barley, rye (Rye grass mosaic virus)
	A. anacardiae Moh.	Cashew
	Abacarus sacchari Channa	Sugarcane
	Acamina coconaciferae Keifer	Coconut
	Acarthrix trymatus Keifer	Coconut
	Acaphylla theae (Watt.)	Tea
	A. indiae Keifer	Tea
	Aculus amaranthi (Cortis)	Amaranthus
	Aculus schlechtendali (Nalepa)	Apple
	Calacarus capsicae Chak and Mond.	Capsicum
	Calcarus carinatus (Green)	Tea
	Calomerus vitis (Pagen.)	Grape
	Cheriacus sulcatus Keifer	Rice (Rice tungro virus)
	Cisaberoptus kenyae	Mango
	Diptilomiopus assamica Keifer	Citrus
	Eriophyes sp.	Arjun[+]
	E. cernuus Masses	Ber[++]
	E. cheriani	Pongamia[++]
	E. ficivorus Channa	Fig
	E. sheldoni Ewing	Citrus

Contd...

Table 2.9–Contd...

Family	Mites	Host Range
	E.vitis (Pagen.)	Grape
	Metaculus mangiferae (Attiah)	Mango
	Phyllocoptruta musae K.	Banana
	P. oleivora (Ashmead)	Citrus
	Phyllocoptes fructiphilus Keifer	Rose (Rose russette virus)
	Oxypleurites convovuli channa	Sweet potato
	Rhyncaphytopus ficifoliae Keifer	Fig
	Floracarus fleshneri keifer	Citrus
Tenuipalpidae		
	Raoiella indica Hirst	Coconut, arreca palm
	Brevipalpus obovatus (Donn.)	Citrus, coffee
	B. calcifornicus (Banks)	Coconut
	B. deleoni	Grapevine
	B. phoenicis (Geijskes)	Coffee, rose, tea, coconut, guava, citrus
	B. obovatus Donn.	Tea
	Dolichotetranychus floridanus	Pineapple
	Larvacarus transitans (Ewing)	Ber
	Tenuipalpus punicae	Pomgranate
Tarsonemidae		
	Hemitarsonemus latus Banks	Jute
	Tarsonemus spinipes	Sugarcane
	T. confuses	Tapoica
	Polyphagotarsonemus latus (Banks)	Mungbean, urdbean, cowpea, brinjal, chilli, cotton, sesame, citrus, marigold, large cardamom, tea
	Steneotarsonemus furcatus De L.	Coconut
	S. spinki Smiley	Rice

++: Major pest; +: Minor pest.

A list of important phytophagous mites belonging to Tetranychidae, Tenuipalpidae, Eriophidae, Tarsonemidae and Tuckrellidae is provided for quick overview of the plant mites (Table 2.9). The synonyms of different species are not considered for the table. They are given in separate rows with their hosts. In some cases, host range varies between 100-200 crops eg. host range of *T. urticae* and *B. phoenicis* varies between 150-200 crops. In such cases, only those crops in which they cause economic damage is given.

Chapter 3

Stored Mites

Mites are the smallest of the stored-product pests. Storage mites include mites belonging to acarid families, predatory and parasitic mites, which are found regularly in stored grains, human and animal food, nutrient ingredients and processed stored products or materials. These are gaining importance due to their increasing incidence and their association/ interaction with fungi and insects causing rapid qualitative and quantitative deterioration of grains. Several species of mites infest stored foods and other organic debris such as grain, flour, cereals, dried fruits and vegetables, pet foods, cheese, dried milk, ham, sugar, tobacco, molds, bird and animal nests, etc. These mites often prefer a moist, damp location (Jeffery, 1976; Hage-Hamsten and Johanson, 1992). Sometimes the surface of infested material appears to move due to the enormous numbers of mites (barely visible to the unaided eye). Heavy infestations of grain mites have a "minty" odor, best detected when mites are crushed between the thumb and forefinger, and held to the nose. Also, a coating or piles of brownish "mite dust" may appear on open shelving, around the base of flour sacks, on the surface of cheese or in other foods. Such piles consist of dead and living mites, cast skins and feaces.

In India, food grains pass through an average of fifteen points between the farmer and the consumer, which enlarges the scope for quantitative and qualitative losses. At each point, mites can enter the food grain ecosystem, which have the potential to turn the grains unpalatable and unfit for human and animal consumption. Under suitable conditions, many of the mite species multiply rapidly to form extremely dense populations and attain pest status.

3.1 Distribution

Several exhaustive reports are available from Czechoslovakia (Zdarkova, 1996), India (Pillai, 1957; Ghai, 1964; 1976; Prasad, 1974; Mathur, 1979; Mathur and Mathur, 1982; Putatunda *et al.*, 1997; Putatunda, 2001; Singh and Gulati, 2001), UK (Hughes, 1976) and USA (Sinha, 1964). In India seventy species of stored product mites have already been reported. The important acarid pest species are *Acarus* sp., *Tyrophagus* sp., *Suidasia* sp., *Glycyphagus* sp., *Lardoglyphus* sp. and *Lepidoglyphus* sp. Table 3.1. In UK based study, 21 per cent of 571 samples were infested with stored mites, which increased to 38 per cent after six months storage in volunteers home. Increase in the percentage of contaminated samples following storage in a domestic situation suggests that the domestic environment is an important factor in developing infestation.

Among the different groups, Astigmata constitutes a dominant group in stored grain ecosystem followed by Prostigmata and Mesostigmata.

3.2 Morphology

Most of the stored product mites are microscopic to minute in size. They are colour less white dull coloured mites, which mostly go unnoticed in large bulk of samples. The important families are Acaridae, Carpoglyphidae, Glycyphagidae and Pyroglyphidae.

3.2.1 Acaridae

Majority of the storage mite fauna belongs to family Acaridae. Mites possess a pallid, soft cuticle varying in colour from pearly white to yellow brown, which helps in respiration. All the species are free living. The gnathosoma, although very mobile, is only partially retractile with in the body. A dorsal propdosomal shield is present. The body setae are usually smooth or sparsely pectinate. Claws are well developed and articulate with a pair of sclerites joined to the end of the tarsus.

3.2.2 Carpoglyphidae

Absence of stigmata, coxal field I is totally closed which is an important taxonomic character of this family

3.2.3 Glycyphagidae

This is another important stored mite family in which males donot possess anal suckers. Females are with genital opening at the level of coxa III and IV or if large at the level of coxa I and II.

3.2.4 Pyroglyphidae

Pyroglyphidae mites belong to the Astigmata so they do not bean stigmata. Vertical setae are absent in both sexes, which is a charactestic feature of the family.

3.3 Commonly Occurring Genera (Putatunda, 2005)

Acarus

Sigma I is more than three times sigma II. In male on leg I, femur is highly enlarged and bears a conical part.

Aleuroglyphus

Chelicerae and legs reddish brown in colour, vertical setae at the same level.

Caloglyphus

All dorsal setae present.

Lardoglyphus

In female claws bified, in males legs III highly modified into two conspicuous process.

Mycetoglyphus

Vertical setae shorter than genu I.

Rhizoglyphus

In Legs I and II tarsi bears enlarged cone.

Tyroborus

Spine like setae on the dorsal apex of tarsus I.

Tyrolichus

Lateral setae are 4-6 times longer than dorsal setae.

Tyrophagus

Needle like setae on the dorsal apex of tarsus I.

Carpoglyphus

Coxal field I totally closed.

Austroglycyphagus

Sigma I and II on genu of leg I is of same length.

Blomia

Two pairs of vertical setae are closed to each other.

Glycyphagus

Crista metopica present sublateral scale absent.

Lepidoglyphus

Sigma II is three times longer than sigma I on genu of leg I.

Euroglyphus

Absence of long posterior terminal setae.

Dermatophagoides

Long terminal setae are present.

3.4 Nature of Damage

Mites have feeding preference for germinal portion of grains. Ground/broken forms are more preferred than whole grains because of many receptor sites on germ portion (Kohli and Mathur, 1994). When given choice, mites generally distribute in all the directions before selecting the more suitable of the two-offered substance (Gulati and Mathur, 1998). Protein-fat combination and concentration or the amount of particular substance plays a role in attraction.

Due to their feeding, colour of grains change from shiny to dull colour, which slowly turn into blackish colour. A layer of powdery mass is also visible during high infestation (Figure 3.1). The thickness of powdery mass depends on the infestation level.

Mites contribute towards the loss in germination (Gulati *et al.*, 1999), nutritive value (Braude *et al.*, 1980), weight and organoleptic properties of grains. Per cent losses in germination due to different mite species were in the range of 20-38 in wheat grains, 5-12 in

Figure 3.1: Damage Caused by Stored Mites

pulses and 35.8 in vegetable seeds after six months of infestation. Low fat acidity value, decrease in protein and starch content and increase in reducing sugars in mite infested grains were observed by many workers. Heavy infestation lead to tainting and minty smell in grains that become unfit for human consumption. Laboratory studies have shown the potential of acarid mites in reducing the weight of grains by 17.3, 20, 26 and 7-18 per cent in groundnut, sugar beet seeds, wheat and cheese, respectively, in six months.

Mites are also associated with storage fungi (Griffiths *et al.*, 1959) and are involved in the initiation and spreading of hotspots. Under laboratory conditions, mites are able to raise the temperature of grain by 10°C. Mixed population of mite, insect and fungi leads to 15°C rise in grain temperature under storage conditions. Complete deterioration of grains occurs at this stage.

3.5 Biology

The life cycle of acarid mite is usually direct, a hexapod larva emerging from an egg, passes through three octapod nymphal stages, *viz.*, proto, deuto and tritonymph, before becoming an adult. In some acarid species deutonymphal stage does not occur and the protonymph moults directly into tritonymph. Each stage is separated by a resting period.

Mating is pre-requisite for egg deposition, which lasts for few hours. During their lifetime, male and female mate many times. Female lays an average of 100-120 eggs in her lifetime. Eggs are round to oval in shape. Hatching, larval and nymphal period varies between 1-2, 1-2 and 4-8 days, respectively. Stored product mites require 7-15 days to complete their life cycle. Generally females live longer than males. Acarid mites have short life span and high reproductive potential *i.e.* several hundreds of progenies can be produced from a single pair.

Under unfavourable conditions, in some acarid species, hypopus stage occurs which can exist for several months without feeding (Griffiths, 1966; Knutte, 1987). It does not move much under its own power but is transported from place to place by clinging to small animal forms such as insects or mice. Most hypopi rely on air currents for distribution. They have been known to survive for at least seven months in dry flour and can withstand lower temperatures than the active form. When the hypopi encounters favorable conditions, it sheds its skin and resumes normal growth and development.

3.6 Storage Mites as Energy Transformers

T. putrescentiae and *R. echinopus* are efficient energy carriers from producer to secondary consumer. 40-60 per cent of the foods consumed by these mites are transferred to body and egg production. Gross production efficiency of growth is generally higher in case of acarid mites (35-36 per cent) than beetles in which it is 10-20 per cent (Sinha, 1964). In different life stages of mites the cummulative net production efficiency value (Production assimilation) ranges from 77-89 per cent whereas in case of beetles the highest net production efficiency is recorded in *Rhizopertha dominica* during late larval stage which is 38 per cent. As a mite consumes energy, it is altering the ecosystem by transferring the energy into biomass, heat of respiration, egested materials and chewed uneaten material which is accumulated and pollute the inert granular space.

3.7 Storage Mites as Source of Allergens

Several stored-product insect and mite species are confirmed as sources of allergens. The most important are storage mites because they cause occupational asthma by inhalation and anaphylactic reactions when ingested in high numbers. Perennial respiratory diseases like chronic rhinopathy or bronchial asthma can often be related to an allergy to indoor allergens (Teho *et al.*, 1985; Musk *et al.*, 1989). Perennial respiratory diseases like chronic rhinopathy or bronchial asthma can also be related to an allergy to indoor allergens (Woodcock and Cunnington, 1980).

Some of the storage mites *e.g. Acarus, Tyrophagus, Glycyphagus, Lepidoglyphus, Cardoglyphus, Gohieria* are important source of allergens for farmers and workers in grain storage/transfer facilities (Cuthbert *et al.*, 1979; Blainey *et al.*, 1989; Iverson *et al.*, 1990; Van Hage-Hamsten *et al.*, 1990). *T. putrescentiae* bodies are the source of 20 antigens, out of which atleast five are allergens (Green and Woodcock, 1978; Wraith *et al.*, 1979).

Allergic reactions due to mites are reported in persons handling the grains. 73 eastern-Polish farmers growing hops and other crops were examined. Tests with storage mites showed positive reactions to *A. siro* in 9.6 per cent, *L. destructor* in 17.8 per cent, and to *T. putrescentiae* in 13.7 per cent (Spiewal *et al.*, 2001). Allergen sensitization, especially to mites and pollens, was significantly associated with asthma and rhinoconjunctivitis in a farming community.

Its severity increases many folds due to its perennial nature and presence of multiple allergens (Revsbech and Anderson, 1987; Revsbech and Ducholam, 1990). Prolonged contact with mite-infested foods may produce a mild dermatitis known as „baker's" or „grocer's itch." Also, if mites are taken internally with infested food, stomach disorders may result. Dermatitis, Bronchial asthma and Chronic rhinopathy are observed in patients associated with mite allergens. Conjunctivitis and Perioral syndrome are also found to be related with stored mite infestation.

3.8 Polluters of Human Food

Mites' presence in the stored product makes the product unacceptable due to minty odour, change in colour and quality of food. Minty smell is associated with lipid secretion, which originates from paired oil glands. Mite attack accelerates the biological oxidation of fats, which causes rancidity. In heavily infested material, brown dust is accumulated, sometimes up to 2 cm depth or more. However, cheese containing thousands of acarid mites are now reported which is responsible for its sharp flavor.

3.9 Bioecology of important stored mites

3.9.1 *Acarus siro* L.

It is commonly called as grain mite (Table 3.1) and one of the most important pest of storage premises and processed cereal products. It occurs in dense populations. As many as 10,000 mites in 200 g wheat sample have been reported. In one study, grain mites comprised 42 per cent of the mites taken from 30 farm granaries in Central Canada (Lyeon). Body of the mite is almost colorless, but the mouthparts and legs are pale yellow to reddish-brown. The adult females are slightly larger than the males. The body of the male is smoothly rounded at the posterior end; that of the female is more oval with the posterior edges slightly indented in the middle. The body has a groove across it, dividing it into two areas. Two pairs of fairly long hairs trail at the end of the body.

At 25-31°C temperature and 80-85 per cent relative humidity, the rate of reproduction is highest in *A. siro*, however, 61 per cent or lower RH is detrimental to grain mites (Cunnington, 1965). Grain mite development is extremely sensitive to humidity levels outside the range of 75 to 85 per cent. It is also known to feed on storage fungi belonging to genera *Aspergillus* and *Eurotium*. A life span of *Acarus* sp. is 13-16 days on wheat germ whereas it is 21 to 32 days on yeast. Synthetic diets containing vitamins and mineral salts increase the mite longevity.

3.9.2 *Tyrophagus putrescentiae* Schrank

It is an important pest of wheat, pulses, millets, groundnut, cheese, mushrooms and other stored products having a high fat or protein content and commonly called as mold mite. It is also recorded from mixed feeds, brewer's yeast, whole-wheat flour; soy flour; wheat germ, cheese, rye bread, white bread, and mixtures of oats and barley. Other known foods include cultivated mushrooms, various seeds, fruits, grain and straw stacks in the field, decaying animal and vegetable matter, onion, bacon, figs, dried milk, cheese, ham, dried bananas and copra.

Males and females (Figure 3.2) are about half mm long and have a small translucent body with almost colorless mouthparts and legs. The body, somewhat slender, bear hairs which are more numerous and longer than those on the grain mite. The fecundity of the mite varies considerably under different conditions of temperature and humidity. The temperature below 10°C and above 30°C is unfavourable for the development of mite although they can survive at these temperatures. On an average 118 eggs are laid by female in her life. Egg hatchability ranges between 90-94 per cent. Under optimum conditions (27±1°C and 80-85per cent RH), duration of egg, larva, protonymph and tritonymph is 4.5, 3.4, 3.2 and 3 days, respectively, on wheat flour. Male and female longevity is 15 and 20 days, respectively. At 20°C the female can live even for 150 days

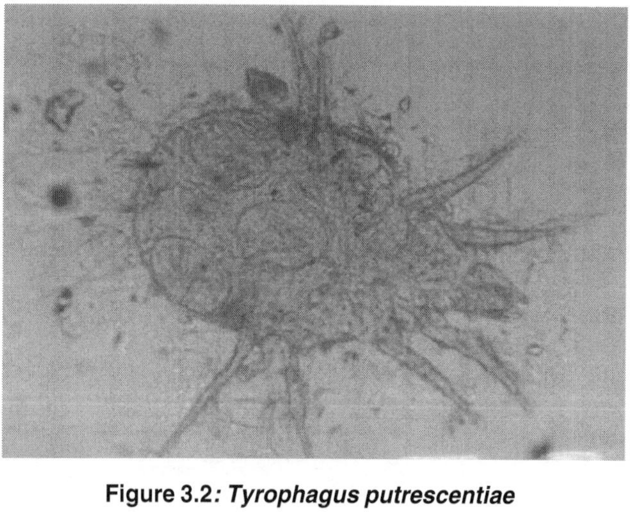

Figure 3.2: ***Tyrophagus putrescentiae***

(Boczek, 1975). Life cycle is completed in 10-14 days. Diet containing methyl p-hydroxy benzoate is best for *T. putrescentiae* as it completed its life cycle within 8-10 days at 25°C and 75per cent RH.

3.9.3 *Tyrophagus caseni*

The cheese mite, known to cause dermatitis, is larger than both the grain mite and the mold mite. It has stout, well tanned, faintly wrinkled legs and tanned mouthparts. Males and females are similar except that females are larger. The life cycle requires 15 to 18 days at the ideal temperature of 73°F and an RH of 87 per cent. No hypopi are formed.

3.9.4 *Suidasia nesbitti* Hughes

This mite is associated with wheat, wheat flour, rice bran, pearlmillet, pulses and other stored grains. Depending upon the food, fecundity ranged between 120-130 eggs per female. Egg, larva, protonymph and tritonymph period ranged between 2.7 to 3.7. 2.6 to 3.6, 2.4 to 3.6 and 2.3 to 3 days, respectively (Singh, 1985). Diet containing wheat germ is preferred by this mite out of various foods tested.

3.9.5 *Glycyphagus* sp.

The characteristic feature of this mite is the length of hair, which is longer than length of body (hairy). It has sharp pointed claws with hairy tufts. Body is cream coloured, small and pear-shaped. It shows jerky movement. It is recorded from grains, flour and cheese. At 23-25°C and 80-85 per cent relative humidity, it completes its life cycle in 20-22 days.

3.9.6 *Carpoglyphus* sp.

It has been found on all kinds of dried fruits, honey combs and food substances containing sugars. It can complete its life span at all temperatures varing from 5 to 35°C, however, egg laying stops below 10°C. The legs are slightly pinkish in colour, which are more slender in females as compared to males.

3.9.7 *Lardoglyphus* sp.

This mite is mainly associated with salted fish and other fish products. *L. konoi* is one of the best-studied species of this genus. The hypopus formation (inactive stage) is higher in nutrient deficient diets and minimal in fish/yeast powder diet.

3.9.8 *Rhizoglyphus* sp.

It is one of the most wide spread genera in stored food/grains and soil fauna. At 25°C and 75 per cent relative humidity, it completes its life cycle in 8-10 days. *R. echinopus* can utilize the lipid content of a number of substrates but are unable to digest the proteinoids *e.g.* collagen, casein, keratin etc.

3.9.9 *Aleuroglyphus* sp.

It is common in wheat, flour, bran, chicken meal and dried fish products where it multiplies and form large population.

3.9.10 *Tarsonemus granius*

It is commonly called as glossy grain mite due to white dot on cream-coloured body. It has very small, shiny and pod-shaped body with a large gap between second and third pairs of legs. The last pair of legs is stick-like in females. It is normally encountered in stored cereals and known as fungus feeder.

3.9.11 Tydeid mite

It is very small, shiny with pod–or diamond-shaped body. Sometimes have hairs on 1st appendages. All legs originate close together and are slightly orange in colour. It is reported from grain residues and known as fungus feeder

3.9.12 *Aeroglyphus robustus*

This warty grain mite has round body. It is comparatively a large mite with arched hairs. Hairs are shorter than body, most being equal in length. It is commonly encountered in farm-stored grain. This mite is reported to feed on seed germ and moulds.

3.9.13 *Androlaelaps casalis*

It is a predatory fast moving mite, which feeds on acarid mites. This large, brown scleratized mite has large legs. First legs are bowed and there are markings on the side of body.

3.9.14 *Blattisocius* sp.

It is a predator of other mites and stored insects. Body is white, plump and pear-shaped with well-developed mouthparts and no hairs on body. Mite body is white in colour and show fast movements.

3.9.15 *Cheyletus eruditus*

This mite predates on other mites and has well-developed pedipalps (= grasping mouthparts). Body is white colour, fairly large in size with legs usually spread out and shows fast movements.

3.10 Management of Storage Mites

Management of stored mites is complicated due their short generation time, high reproductive potential, broad food niche, interactions with other pests and pathogens and unique adaptations for dispersal. Historically, control of these acarine pests has relied on the use of synthetic miticides and insecticides, but this option is now limited due to documented resistance and withdrawal of registration of some products. Alternative control strategies, including cultural and biological control, have shown limited success, but need to be further developed and implemented.

3.10.1 Preventive Measures

Unfortunately, once large numbers of mites are noted, control becomes very difficult. Contamination during storage must be avoided. Grains should be stored below 12 per cent moisture content of grains. Low moisture content is critical for mite control than low temperatures. At temperatures below 5°C most storage insects die whereas, mites merely stop breeding. Cleansing and disinfections of stores, preventing breaking of packaging seals are some of the other factors, which should be taken into consideration before storing the grains or food products. Foods should never be stored under damp, poorly ventilated conditions. If necessary, air circulation should be increased to reduce the relative humidity.

If bins or large containers are used for storage, do not dump new replacement foods (flour, grains, etc.) on older unused foods. Allow the original contents to become used up or exhausted, if possible because unused materials may become damp, moldy over time and act as source on initial infestation. Avoid prolonged storage. Stored foods should be placed in containers with tight-fitting lids, ideally screw type. Flour or packaged foods should be stacked on pallets to permit air ventilation and to prevent possible floor dampness from penetrating sacks or packages. When products become infested with mites, locate the source of infestation and eliminate it. Remove any bird or rodent nests near the storage area.

3.10.2 Use of Fumigants

Fumigants are found to be ineffective on eggs and early stage larvae, which result in resurgence of mite population after few days. Moreover, mites need 1.5 times higher dose than recommended for insects. Spraying of lindane-malathion or Chlorpyriphos-phoxim mixture, etrimphos and primiphos methyl are reported to be effective against *A. siro, T. putrescentiae* and *G. destructor*. Bin walls treated with malathion provide a protective barrier against mite pests.

3.10.3. Use of Inert Dust

Innert dust @ 3 and 5 g/kg provide complete control against *A. siro* and *G. destructor* at 10°C and 14.5 per cent moisture content after 28 days of exposure (Cook and Armatage, 1999).

3.10.4 Use of Botanicals

Studies on biodegradable botanicals showed that spice oils (black pepper, mint, turmeric, garlic) have inhibitory/repellant activity against stored mites (Potts and Rodriquez, 1979; Gulati and Mathur, 1995; Gulati, 2002). Neem oil (Gulati, 1998), vegetables oils (Sesame, mustard, groundnut) and Tulsi, Mahua extract also provide effective protection against acarid species. Degree of effectiveness depends on the type of species and nature of product.

3.10.5 Use of Photoperiod

Photoperiod also exerts a powerful influence on their life processes as they generally remain in dark during post harvest storage. Increase in light exposure leads to decrease in fecundity, egg viability, growth and increase in hatching and life history durations (Kohli and Mathur, 1993).

3.10.6 Pheromones as Repellants

T. putrescentiae contains neryl formate in its body fluids which acts as an alarm pheromone when released. Citral, a closely related compound, serves as the alarm pheromone of four other astigmatid mite species. Potts and Rodriguez (1979) reviewed techniques for protecting stored foods with natural mite allomones. Filter paper discs impregnated with these oils were placed adjacent to a food product inside the food wrapper. This treatment provided complete protection when applied at a concentration of 0.5 per cent.

Table 3.1: Stored Product Mites Associated with Stored Products/ Grains

Family	*Mite-pest*	*Store Product*
Acaridae	*Aleuroglyphus ovatus*	Wheat flour, cotton seed, Coconut, eliachi powder, suji
	Acarus sp.	Wheat flour, wheat dalia, pea, rice, cotton
	A. farris	Wheat flour
	A. siro	Wheat flour
	Caloglyphus sp.	Wheat flour, gram, suji
	C. berlesei	Wheat flour
	C. mycophagus	Wheat flour
	C. rhizoglyphoides	Wheat flour
	Mycetoglyphus sp.	Stored grains
	Rhizoglyphus sp.	Wheat flour
	Tyrophagus longior	Wheat flour, gram, dhania, debris of shops
	T palmarnum	Wheat
	T. putrescentiae	Wheat flour, pulses
	Thyreophagus entomophagous	Wheat, wheat flour, gram, oat, sugar, animal feed, debris of shops, bajra
	Tyroborus lini	Stored grains
	Tyrolichus casei	Stored grains
Chortoglyphidae	*Suidasia nesbitti*	Wheat flour in mills, bajra
Glycyphagidae	*Austroglycyphagus geniculatus*	Rice, animal feed capsules
	Glycyphagus sp.	Wheat flour
	G. domesticus	Wheat
	Lepidoglyphus destructor	Wheat, wheat flour
	Blomia sp.	Wheat flour
Pyroglyphidae	*Dermatophagoides farinae*	Wheat flour
	Euroglyphus sp.	Wheat flour
Urodinychidae	*Leiodinychus parasiticus*	Gram, wheat flour, dry coconut, jeera
Carpoglypidae	*Carpoglyphus* sp.	Wheat
Hisiostomidae	*Rhizoglyphus sp., R. robini*	Bulbs
	Lardoglyphus konoi	Fish products

Table 3.2: Predatory/Parasitic Mites Associated with Stored Products

Cheyletidae	
Acaropsis sollers	Gram, gram dal, bajra, mustard, oat, rice
A. docta	Gram dal, urd dal, wheat
Cheletopsis major	Gram dal
Cheletonella sp.	Wheat, rice
C. sugosa	Wheat, rice
Cheletonata sp.	Wheat
Chelonotus sp.	Wheat
Cheletomorpha sp.	Wheat
Cheyletus sp.	Wheat dalia, debris of shops
C. fortis	Mango, powder, coffee
C. malaccensis	Bajra, rice, wheat, gram, barley, wheat flour, mango powder, mustard debris
C. malayensis	Rice, gram, debris of shops
C. strenuus	Urad dal
Neoacaropsis granulatus	Wheat
Eucheyletia sp.	Wheat
Eucheletopsis sp.	Wheat, gram dal
Hemicheyletia sp.	Wheat
Stigmaeidae	
Eustigaeus sp.	Wheat, dry coconut, debris of shops
Indostigmaeus sp.	Wheat
Tydeidae	
Pronematus sp.	Kabuli gram, pea, rice paper, wheat dalia
Acarophenaxidae	
Acarophenax nidicolus	Wheat
A. tribolii	Wheat
Pygmethoridae	
Pygmethoride sp.	Debris of shops
Cunaxidae	
Cunaxa sp.	Wheat

Contd...

Table 3.2–Contd...

Pyemotidae	
Pyemotus sp.	Wheat, pulses
Urodinychidae	
Leiodinychus krameri	Stored products
L.parasiticus	Stored products
Laelapidae	
Androlaelaps casalis	Stored products
Hypoaspis aculifera	Stored products
Proctolaelaps sp.	Stored products
Ameroseiidae	
Klemania plumigera	Stored products
K. plumosus	Stored products
Ascidae	
Arctoseius butleri	Stored products
Blattisocius sp.	Stored products
B. keegani	Stored products
B. tarsalis	Stored products
Melicheres sp.	Stored products

3.10.7 Use of HTST Technique

Infested grains are treated to an appropriate lethal temperature/ time combination. Research on high temperature control has focused on using hot air in a fluid bed, pneumatic conveyer equipment and on the use of high frequency waves, microwaves and infra-red radiations to heat the grain/ flour before rapidly cooling the commodity. High temperature of 300-350 °C for 6 seconds was found to kill mixed population of acarid species in grains (Mourier and Poulsen, 2000). At this temperature/time combination, reduction in moisture content was 0.5-1 per cent and no adverse effect on the functional properties of seeds.

3.10.8 Use of Biocontrol Agents

Biological control of mites is being exploited using several predatory mites *viz. Chelytus eruditus, C. malaccensis, A. docta, Androlaelaps casalis, Blattisocius* sp. *P. tritici* which are known to feed on acarid mites. These agents belong to four main families *viz.* Ascidae, Acarophenaxidae, Cheyletidae and Pyemotidae (Table 3.2).

3.10.8.1 Ascidae

Stigma between IIIrd and IVth legs with straight peritremes, metasternal shield small, female genital shield not flasked shape and dorsal shield with more than 23 pairs of setae. Majority of the mites species belonging to this family feed on insects and nematodes and are dealt in relevant chapters. Among the acarines, *Arctoseius cetratus* (Sellnick) limit the population of Mushroom pest, *Tarsonemus myceliophagus* Hussey. *Lasioseius parberlesi* Tseng is a natural enemy of rice pest, *Steneotrasonemus spinki* Smiley (Tseng, 1984) and hold its numbers at non-injurious level.

3.10.8.2 Acarophenaxidae

Leg I modified into a club like form with claw. Anterior propodosoma projects over the gnathosoma. The genus *Acarophenax* is mainly associated with stored grain pests.

3.10.8.3 Cheyletidae

They are white, yellow or orange colour mites with prominent thumb-claw complex. Chelicerae and rostrum fused into a cone, peritremes arched or forming an M-shaped configuration. *Cheyletus eruditus* (Schrank) is a biocontrol agent, which is currently being used for *Acarus* spp. (Pulpan and Verner, 1965; Zdarkova, 1986). It completes its life cycle in four weeks and prefers slightly higher temperature and lower humidities than the prey mites (Solomon, 1969). It can survive long periods of food scarcity but in case of choice, prefers immature stages of prey (Berreen, 1984). It has a natural tolerance to certain organophosphate insecticides (Zdarkova and Horak, 1987). Release of one predator/ 20-50 pests in oilseeds and 2000 predators/100 m^2 in empty store houses get full protection from acarid mite population.

3.10.8.4 Pyemotidae

They are milky white, spindle shaped and rounded segmented mites. Female possess pseudo stigmatic organs. Leg IV of female with pretarsus, claw and pulvillus but without apical whip like setae, idiosoma is elongated, gravid female with sac like hysterosoma. These parastic mites have a reduced life cycle. Mated females seek out hosts and attach to the host, sometimes paralyzing the hosts with an injected toxin. Some species of genus *Pyemotes* casue dermatitis in humans. However, work on biosystematic research has shown that genus *Pyemotes* contains two groups of species (Moser *et al.*, 1987); one is parasitic only on scolyted beetles and are venomless, the other is venomous, polyphagous and very lethal to its prey but of reduced dermatological risks to humans. Some species are discussed in insect control chapter.

Chapter 4

Soil Mites

The soil is a unique habitat that supports rich and diverse life. It is a matrix of myriad solid mineral particles and pores filled by water, air, and decomposing organic matter. Most soil animals absorb and lose water through their integument and depend on water-saturated atmosphere for their existence. If the soil dries out, the animals in soil either die, exist as resistant cysts or eggs, or migrate away from unfavourable conditions. The community of soil invertebrates is one of the most species rich communities in terrestrial ecosystem whose interactions influence the crop productivity. Acarines constitute an important biotic fauna among the soil arthropods. In India, among soil mites, the number of species/genus/ families known under Prostigmata, Mesostigmata, Cryptostigmata are 12/10/4, 150/55/18 and 425/185/55, respectively (Gupta and Gupta, 1999). In the same year, Haq also reported 29/21/18 cryptostigmatid mites from Kerala (Haq, *et al.* 1999).

4.1 Oribatid Mites

Oribatid mites belong to the suborder Cryptostigmata (Oribatida). Since these mites are dark coloured and covered with a rigid exoskeleton, they are popularly known as 'beetle' or "moss" mites and are the world's most numerous arthropods living in soil. Their intricate relationship with in soil ecosystem and the roles played by individual species remain 'terra incognita'. Their density in forest soil can reach hundreds of thousands of individuals per square metre. There are several thousand described species, yet the fauna of much of the tropics is still unknown. Adult oribatid mites usually have strong exoskeleton, hardened by sclerotization as in other mites, and by mineralization, similar to millipedes and isopodes. These slow moving mites are 0.2–1.0 mm in length and occur in the top layer of soil, litter debris, sometimes also on plants, mosses and lichens. Phoresy is a wide spread phenomenon in the oribatids which has been reviewed earlier

4.1.1 Morphology

The gnathosoma can usually be withdrawn into the body and is concealed by the overhanging rostral region of the propodosoma. The chelicerae are prominent and the pedipalp has five free segments. In addition, the infra capitulum bears a pair of prominent toothed structure (rutella) which assist in breaking up of food. Lamellae and pteromorphae arise from the idiosomal portion of the body. The lamellae are paired lateral folds of the propodosoma, terminating in a lamellar seta. The pteromorphae are wing like projections from antereo-lateral margins of hystereosoma.

4.1.2 Biology

All the oribatids have a calyptostastic prelarva but the larva, the three nymphal stages and the adult are normal. The regressive prelarvae are usually ovoid and lacks appendages whereas non-regressive prelarvae either have well developed appendages or show vestiges of appendages but lacks segmentation.

Sexual dimorphism is strongly demonstrated in this group where bisexual forms are numerous with male and female are in about equal proportion. However, many species are obligately parthenogenetic, with no males recorded. In contrast to many other microarthropods, oribatid mites reproduce slowly. In cold climates, their life cycle can stretch up to seven years (usually one-two years). Females lay few eggs. With their slow metabolic rate, slow development and low fecundity, oribatids are not capable of fast population growth and are usually restricted to relatively stable environment.

Oribatid mites graze on fungi and algae, consume decomposing plant matter and occasionally dead collembolans. Some oribatids feed on live nematodes. For many groups feeding habits are still unknown. Adult oribatid mites experience relatively little predation due to their strong exoskeleton, though they are subject to predation by ants, certain beetles and small salamanders. Many soil predators attack the soft-bodied nymphs of oribatids. Oribatida comprise an important component of soil decomposers, their abundance, species composition and diversity in a particular habitat serve as good indicators of soil "health".

4.1.3 Life Cycle

The life cycle consists of egg, larva, three nymphal stages and adult. In many cases, the larva and nymphs are markedly different from the adults. A female lays an egg, which changes into prelarva before moulting to a larval stage. Three nymphal stages *viz.*, protonymph, deutonymph and tritonymph occur in the life cycle before attaining maturity as an adult.

4.1.4 Decomposition and Mineralization

The soil-dwelling mites contribute to plant litter decomposition, soil formation and nutrient cycling. They feed on fungi and nematodes and are extremely important organisms in maintaining soil health and fertility. The movement of mites in the soil may also mix the soil up so that microbes can have greater access to residue. Mites can introduce microbes to fresh residues so that decomposition proceeds. Their grazing on microbes can stimulate microbial activity. Studies have shown that when mites are excluded from residue, its decomposition rate is slowed by an average of 23 per cent. Researches have recently demonstrated that mites are integrally linked with the nitrogen cycle.

These are primarily the inhabitants of the upper layers of soil and surface litter. Evidence also shows them as predators of nematodes and vectors of cestodes. These well-sclerotized mites are beneficial in nature as they are believed to accelerate the decomposition of organic matter and nutrient release in forestry plantations. They are reported to increase the litter decomposition rate by 38 per cent in *Eucalyptus* foliage to 49 per cent in *Oak* foliage. Decomposition (decrease in the mass of substrate) results from leaching of soluble materials (intracellular components), catabolism/ oxidation of organic matter by converting fixed carbon into CO_2, water, energy and comminution of litter.

In soil, oribatid mites constitute more than 50 per cent of the total micro-arthropod population. Their densities range from 50,000 m^{-2} in tropical to 300,000 m^{-2} in boreal forests where large amount of genetic organic matter occurs.

Oribatid mites have well developed mouthparts, which are capable of fragmenting organic matter. They also selectively feed on microflora adhering to this detritus. Due to fragmentation and comminution (reduction to small fine particles), new surface area is created for microbial colonization (Figure 4.1). Changes in surface area of substrate and repositioning of fragments increases the leaching and oxidation rates. Therefore, mites enhance the decomposition rate directly by contributing to weight loss of the

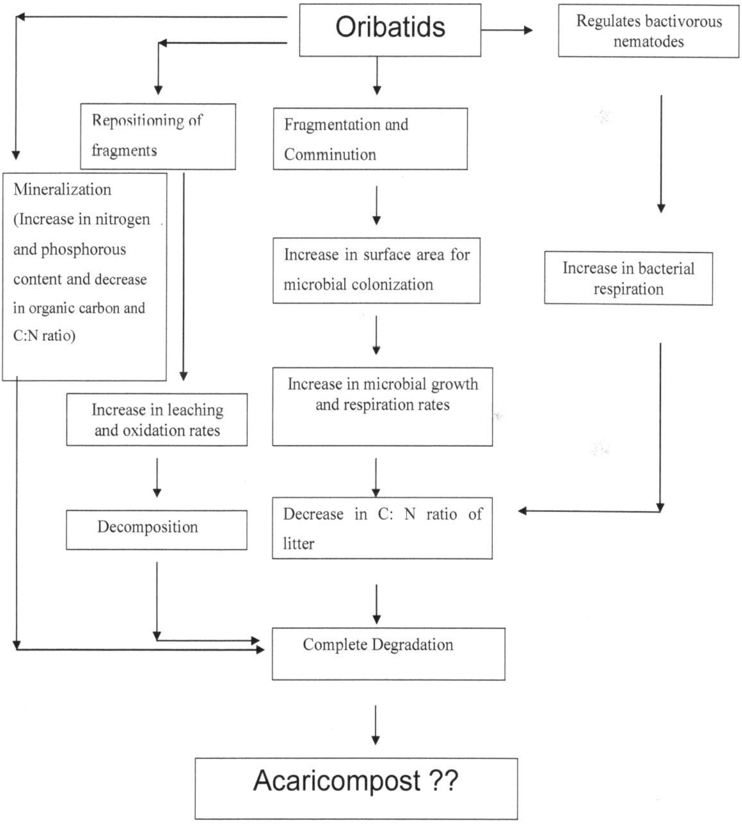

Figure 4.1: Oribatid Mite Activity in Soil

litter and indirectly by spreading microflora for fast microbial degradation activity. Many investigators have found a positive correlation of oribatid mites with organic carbon, available nitrogen and soil moisture. In litter, maximum population is recorded in monsoon and autumn by various workers. Soil acts as overwintering ground for oribatid mites where they prefer to remain in the upper 0-8 cm layer of soil.

There are studies, which indicate decline in mite population after pesticide application, thereby, decline in leaf litter decomposition and subsequently poor nutrient release in soils.

4.1.5 Bioindicators

They play a prominent role as bioindicator of soil health. A number of studies on the impact of sheep and cattle grazing on soil mesofauna have shown that mite species richness and densities decrease in soil under significant grazing pressure The soil mites can be good indicators of soil health, as well as indicators of disturbance. They may also have potential as early-warning indicators of alterations to soil structure because they live in the macro-pores and root channels of soil. It is shown that mite abundance and diversity were greater in soils with perennial grass cover than in soils with bare ground and annual grass cover. In hot dry environments they are also the dominant soil animals comprising 70 to 90 per cent of the mesofauna (medium-sized soil animals from 0.2 mm to 3 mm long).

4.1.6 Oribatid Families and Genera

Important families associated with oribatid mites are Brachychthoniidae (*Brachychthonius bimaculatus, Liochthonius* sp., *L. lapponicus, Sellnickochthonius immaculatus, Verachthonius montanus),* Cosmochthoniidae (*Cosmochthonius* sp.), Camisiidae (*Camisia biurus, C. horrida, Camisia* sp.), Trhypochthoniidae (*Trhypochthonius tectorum, Mucronothrus nasalis),* Gymnodamaeidae (*Gymnodamaeus bicostatus, Gymnodamaeus sp, Odontodamaeus sp., Joshuella sp. Nortonella gildersleeveae), Damaeidae (Belba sp., Epidamaeus sp., Dyobelba sp., Quatrobelba Montana),* Eremaeidae (*Eremaeus boreomontanus, E. kananaskis, E. occidentalis, E. plumosus, E. salish, E. translamellatus, E. aysineep, Eueremaeus chiatous, E. foveolatus, E. marshalli, E. masinasin, E. michaeli, E. osoyoosensis, E. tetrosus),* Megeremaeidae (*Megeremaeus kootenai),* Tenuialidae (*Hafenferrefia sp.),* Liacaridae (*Dorycranosus sp.),*

Peloppiidae (*Ceratoppia bipilis, C. quadridentata arctica, Ceratoppia sp.)*, Carabodidae (*Carabodes colorado, C. dickinsoni, C. wonalancetanus, Carabodes sp.)*, Tectocepheidae (*Tectocepheus velatus, Tectocepheus sp.)*, Oppiidae (*Microppia simplissimus, Oppia sp., Oppiella washburni, O. nova, Oppiella* sp.) Quadroppiidae (*Quadroppia ferrumequina*), Suctobelbidae (*Suctobelba* sp., *Suctobelbella* sp.), Caleremaeidae (*Veloppia kananaskis)*, Cymbaeremaeidae (*Ametroproctus tuberculosus, A. reticulates, A. canningsi, Scapuleremaeus kobauensis)*, Passalozetidae (*Passalozetes californicus)*, Scutoverticidae (*Exochocepheus eremitus)*, Scheloribatidae (*Liebstadia similes, Scheloribates pallidulus, Scheloribates* sp.*)*, Oribatulidae (*Oribatula* sp., *Zygoribatula* sp.), Ceratozetidae (*Ceratozetes cuspidatus, C. gracilis, C. kananaskis, C. oresbios, C. watertonensis, C. thienemanni, Dentizetes rudentiger)*, Mycobatidae (*Mycobates altus, M. azaleas, M. brevilamellatus, M. dryas, M. incurvatus, M. punctatus*), Phenopelopiidae (*Propelops canadensis, P. pinicus, Propelops* sp.), Oribatellidae (*Oribatella* sp.), Tegoribatidae (*Lepidozetes sp.)*, Achipteriidae (*Parachipteria nivalis)*

4.1.6.1 Eremaeoidea

Species in this superfamily primarily are found in dry habitats where they live in organic litter. They are also found on trunks and in the canopy of trees, where they feed on the fungal fruiting bodies and the fungal component of lichens. In the soil profile, species of Eremaeidae are found primarily in the litter layer, or in moss and lichens on the soil surface. Occasionally a large percentage of the population may be found in the fermentation layer. *Eremaeus* spp. is the predominant genus associated with this family.

4.1.6.2 Carabodoidea

Many species in the carabodoid genus *Carabodes* are more common on tree trunks and in the crown of trees than in litter. They form part of the diverse, but poorly known, arboreal oribatid fauna which includes species in other superfamilies, *e.g.*, Crotonioidea and Oripodoidea. *Carabodes* species feed on fungi, especially bracket fungi, and lichens on trunks and branches of trees.

4.1.6.3 Cymbaeremaeoidea

Species in the cymbaeremaeid genus *Ametroproctus* are restricted to dry, primarily alpine habitats. These species are probably fungivorous.

4.1.6.4 Licneremaeoidea

The scutoverticid species *Exechocepheus eremitus* Woolley and Higgins is among the dominant arthropods and often the dominant oribatid mite in grass habitats. Occurrence of this genus is so far restricted to dry to arid grassland habitats.

4.1.6.5 Ceratozetoidea

This is diverse superfamily where members of the Ceratozetidae and Mycobatidae, which prefer dry habitats, are dominating fauna. Species of *Ceratozetes, Dentizetes* and *Mycobates* are macrophytophages, microphytophages, fungivores and some feed on carrion. Species of Ceratozetidae are among the oribatid mites of veterinary importance as intermediate hosts and vectors of tapeworms, and could play a similar role as intermediate hosts for tapeworms of Mountain goats.

4.2 Mesostigmatid Mites

The mites belonging to order mesostigmata have adopted themselves to a wide range of habitat. Majority of them are predators and form an important part of the fauna of surface litter soils where they feed on other sluggist arthropods, fungi and nematodes. Phytoseiid mites feed on phytophagous mites and destructive insects, which has already been discussed in previous chapter on plant mites. Parasitism is also wide spread in mesostigmated mites, showing various degrees of intimacy with their hosts. Many species are associated with stored food, carried there on clothes or by rodents and insects. Some of them act as coprophagous, mycophagous or scavenger in nature.

4.2.1 Morphology

The mites of this order range in size from 200 to 2500 µm. Many of the smaller forms, particularly euedaphic species are weakly scleratized and pale in colour but generally the idiosoma is partially or entirely covered by a number of chestnut brown shields separated by striated cuticle. In mesostigmated mites, stigmata is easily seen which is usually situated on the sides of the idiosoma and associated with a tubular peritreme.

4.2.2 Life Cycle

The life cycle consists of an egg, hexapod larva, two nymphal stages and adult, which are typical of this taxon. The resting period between each active stage is of few hours only. In many species, deutonymph is a dispersal stage, which may be prolonged. A larva is hatched from an egg, which after few days of feeding moult into protonymph and then deutonymph. After a quiescent phase nymphal stage finally moults into an adult.

Among the soil borne acarines, mesostigmatid mites are dominant group in horticultural crops and second largest group in field and vegetable crops (Walia and Mathur, 1997). Laakso and Setala (1999) reported that mesostigmatid mites could reduce the biomass of bacterial and fungal feeding nematodes by more than 50 per cent.

Mesostigmatid mites found in soil are mostly nematophagous in nature. They have the potential to reduce the plant parasitic nematode population in soil significantly, which has been proved experimentally by various workers all over the world. Their ability to undergo hypopus stage under unfavourable conditions stimulated their efficacy in suppressing the nematodes. These mites actively search for nematodes in soil and prefer them to other diets. Most of them show high fecundity on nematode diet. A list of mites associated with nematodes has been given in Table 15.1.

4.3 Prostigmatid Mites

The order prostigmata includes a large number of diverse forms, which are weakly sclerotized and vary greatly in structure and biology. Size of the mites varies from 100 to 1600 µm. A pair of stigmata lies at the base of the gnathosoma. Prostigmatids are the dominant group in our soils. They are usually soft bodied and different species have a great variety of feeding habits. Over 14,000 prostigmatid species have been identified throughout the world.

Many species live by sucking the juices of plants (Tetranychidae) and animals (Cheyletidae). Sometimes, only the larva is parasitic in habits *e.g.* Trombiculid larvae suck the tissue fluid of arthropods or water mites which prey on aquatic insects and fresh water mussels. Some species are endoparasitic in nature like demodicid and syringophilid mites spent their life in the hair follicles and sebaceous glands of mammals and the feather quilts of birds.

In the present chapter, among the prostigmatid mites only nematophagous and trombidiid mites are discussed which are found in soil, others are dealt with in separate relevant chapters.

4.3.1 *Alicorhagia fragilis*

It is a prostigmatid mite, which spins a silken thread from its oral cavity and weaves a platform on which eggs are laid. Only female are observed in field and laboratory cultures. All active stages including larvae are aggressive predators of nematodes. Developmental time (egg to adult) is about 21 days at 23°C (Walter, 1988b).

4.3.2 *Alycus roseus*

Among the prostigmatid mites it is a sexually reproducing species. All active stages (larva, protonymphs, deutonymph, tritonymph, adult) are predators of nematodes. One larva can consume 12 nematodes (400 µm in length) in 30 minutes (Walter, 1988b). Developmental time is about 35 days at 23°C.

4.3.3 *Trombidiid mites*

This group is one of the most fascinating groups of mites on which very limited research is carried out. The mites are large, red velvet mites commonly found in soil, litter and other terrestrial habitats. In the larval stage, most mites of the family trombidiidae are ectoparasites on arthropods whereas; adults and deutonymphs are predatory in nature.

4.3.3.1 Life Cycle

The sequence of life cycle stages is similar to that of other prostigmatid mites: egg, prelarva, larva, protonymph, deutonymph, tritonymph and adult. Prelarvae, protonymphs and tritonymphs are calyptostatic in nature.

Eggs are laid in the soil and hatch after 10-30 days, depending on the environmental conditions. Fecundity in the Trombidiidae varies between species and ranges from 60 eggs in *Podothrombium* to 100,000 eggs in *Dinothrombium* sp. (Newell, 1979). The occurrence of the larvae is generally in synchrony with their hosts. They attach to their arthropod host for 1-2 weeks, then detach themselves and move into the soil. Deutonymphs emerge in the summer or autumn and forage on the soil surface and plants. Adults emerge in the autumn. The development and survival of each life cycle stage is affected by the temperature, relative humidity (RH) and food availability/ quality. The optimal condition for postembryonic development is

14-25°C and 80-100 per cent RH. The life cycle is completed in 6-10 months (Robaux, 1974).

4.3.3.2 *Allothrombium* sp.

Eggs are laid in autumn (Dong *et al.*, 1996) and they prefer to lay eggs in wet soil with water content of 12.5 per cent (Zhang *et al.*, 1993). At 7-30°C, oviposition period lasts for 30 days and 300-1250 eggs are laid per female.

With in this genus, some species are restricted in host range (*e.g. A. pulvinum* on aphids) whereas others may be broadly polyphagous (*e.g. A. ovatum* on Homoptera, Lepidoptera and Coleoptera) (Zhang, 1991; Zhang and Saboori, 1996). *A. pulvinum* prefer large aphids and suck their body fluid through wound. They mainly attached themselves on the thorax region of host. *A. pulvinum* can kill their host *Aphis fabae* in three days when the mite load is two or more. One mite/host decreases the reproductive rate of adult host and arrest the development of nymphs (Zhang and Xin, 1989).

Their adults are generalized predators and feed on most of the small arthropods they can handle. They are reported to feed on spider mites, spring cankerworm (*Paleacrita vernata*), cabbage moth (*Mamestra brassicae*), leafhopper (*Rhytidodus decimusquartus*), maize pest (*Zyginidia pullula*) and poplar lace bug (*Hegesidemus habus*). Adults are active on the soil surface and plants during the day but hide in the soil during night and in winter (Zhang and Xin, 1989). *A. pulvinum* adults can consume 7 newly hatched aphids and 85 poplar lace bug per day. It also accounts for 57 and 80 per cent of the natural enemies in orchards and cotton fields, respectively, in China (Chen *et al.*, 1994).

4.3.3.3 *Trombidium* sp.

One female of *Trombidium* sp. lays upto 2000 eggs in her lifetime. Larvae are associated with insects of Lepidoptera, Coleoptera, Diptera, Orthoptera, Hemiptera, Homoptera and Hymemoptera. 2-3 larvae/*Cerotome* beetles, 8 larvae/ *Phyllotreta* beetle, 15 larvae /moth and 41 larvae /housefly are recorded by various workers. Adults feed exclusively on the eggs of collembolans. They spend most of their time in the soil and are only occassionally found on the surface (Zhang and Xin, 1989).

Chapter 5

House Dust Mites

The house dust mites are prevalent in human dwellings in humid geographical areas throughout the world and maintain their body water, which is approximately 70-75 per cent of their body weight. The house dust mites thrive in areas of heavy human traffic where shed human skin scales are abundant as food. They also live in the fine layer of minute dust particles that continually settles on household items. They are found almost world wide but their numbers are considerably reduced at high altitudes and dry climates. House dust mites have been known to be associated with allergies since the 1960's and have become a focus over the years for their involvement with respiratory ailments. They do not bite or sting but harbour strong allergens in their bodies as well as in their secretions, excreta and shed skins. Constant contact with these allergens can trigger respiratory and dermatological complaints in some humans.

As relative humidity influences their population dynamics, their number are large during humid months than dry months (Arlian and Veselica, 1979). Under high humidity levels and constant warm temperatures, mite populations explode, although there are some seasonal fluctuations. The mites are known to concentrate on certain furniture items, especially beds, upholstered lounges, chairs, and in carpets with long fibres. They are attracted to these areas for food, and they feed on shed human skin scales and secretions, house dust, fungal spores, pollen grains, plant fibres and insect scales.

Dust mites dislike strong light and will take harbourage in the seams, ledges, and framework of furniture when exposed. Due to their light weight, dust mites can become airborne during activities such as bed making, and in this way can be distributed throughout the room or house. They prefer a high humidity (80 per cent) and a temperature (25°C), and so are particularly common in beds.

They extract adequate amounts of water from water vapour of unsaturated air and have large surface to volume ratio, which favours rapid evaporative loss of water. However, these mites exhibit several adaptations to conserve sufficient water to compensate for the loss (Arlian and Wharton, 1974):

1. They have impermeable integument (exoskeleton) that encases the body and impedes the evaporative water loss.
2. They obtain water from unsaturated air as water vapour.
3. They lack on organized respiratory system and associated opening, therefore, little water is lost by ventilation.

Water loss is generally through egg production, by evaporation from general body surface, body secretions and through excretion. During dry periods, they form desiccation resistant protonymphal stage, which can survive for months at relative humidity below the critical humidity (0per cent RH) (Arlian, 1989; 1992) for active stages. Clustering behaviour of males (Glass *et al.*, 1998), protonymphs and females (Arlian *et al.*, 1983) is considered as an adaptive behaviour, which helped the mites to reduce water loss and show better defense. Cluster formation restricts water loss passively and thus may save energy that would be used for active water vapour uptake. The clustered mites form a microcosm (Figure 5.1) which effectively decrease the surface area from where water is lost while raising the relative humidity for individuals on the inside of the cluster.

Fain *et al.* (1988) provided a key for all known species of the globally distributed family pyroglyphidae.

According to new concept, house dust mites are included in 'Domestic mites' category alongwith storage mites. As most of the domestic species live in the nests of wild birds, it suggests their nidicolous origin (Fain *et al.*, 1988). House dust mites, in particular includes the species of nest dwelling mite family, pyrolyphidae which have repeatedly been found to occur in the human 'nest' *e.g.*

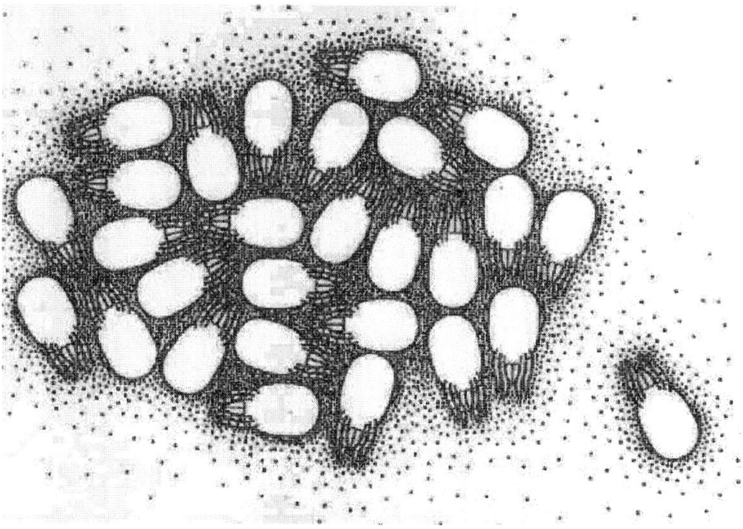

Figure 5.1: Clustering Behaviour by House Dust Mites (Glass *et al.*, 1998)

Dermatophagoides pteronyssinus, D. farinae and *Euroglyphus maynei*. Some storage mites are also reported from house dust, farm dust and other occupational environments *e.g. Acarus siro, Tyrophagus putresentiae, Lepidoglyphus destructor, Blomia kulagini* etc. (Miyamoto *et al.*, 1969; Dar and Gupta, 1979; Kumud *et al.*, 1988; Singh and Gulati, 1998).

5.1 Distribution

Distribution of house dust mites is worldwide. Fains (1967) did the pioneering work on house dust mites and parasitic mites and gave descriptions of 2240 species. Inside the house, they cover virtually everything in the interior of the house-floors, equipments, furniture, shelves, textile objects, beds, upholstory, carpets etc (Cunnington, 1967). Mite growth is affected by temperature, humidity, supplement of food (human dander) and place suitable for growth (Natuhara, 1989). Optimal conditions for their development are recorded as 25°C and 75-80 per cent relative humidity. Among house dust allergens pyroglyphid mites are major source of allergens in house dust. Mite derived allergy because of its perennial nature is considered serious than other aeroallergens causing seasonal allergy.

5.2 Morphology

Majority of the pyroglyphid species are commensals or parasites of birds and mammals; some are free living and occur in the nests of birds and human habitations. The body is often weakly sclerotized. In many males (*e.g. D. pteronyssinus*), legs III are conspicuously larger than legs IV. A characteristic feature of the family is the apical position of solenidion w_1, as well as the famulus and w_2, on tarsus I and the presence of cuticular hooks at the tips of tarsi I and II.

5.3 Biology

D. pteronyssinus, D. farinae and *E. maynei* are the predominant species of pyroglyphidae which occur in house dust, so these are discussed in detail. After copulation, female deposits eggs, which forms prelarvae and hatched as hexapod larva. Two nymphal stages *viz.* protonymph and tritonymph occur in development before moulting into adult. Few scientists have reported only one nymphal stage *i.e.* tritonymph. In between the two stages, quiescent phase also occurs in which mite becomes motionless; the duration of each quiescent phase lasts from few hours to few days under optimal conditions. Life cycle is completed between 19-30 days depending upon the temperature and humidity.

5.4 Mating Behaviour

Males are able to detect with tritonymphs, which will moult into females possibly due to the expulsion of pheromones from lateral opisthosomal glands of tritonymphs and will mount them in advance of the final moult so that copulation can take place shortly afterwards. Mites remain in couple for between 24-48 hours.

5.5 Mites as Source of Allergens

Among pyroglyphids, *D. pteronyssinus* and *D. farinae* are considered to be source of multiple allergens causing allergenicity in humans. Allergenic extracts prepared from live mites, dead mites and fecal material, indicated that the body and feacal extract contain some common antigens (Arlian, 1991). Most of the antigens in feaces extract are also present in body extracts; some antigens although present in body extract are more concentrated in feaces extract and some antigens of body extract are about from feaces extract. Medical conditions that are attributable to the presence of dust mites include allergic rhinitis, asthma, childhood eczema and other allergic conditions (Platts-Mills *et al.*, 1982). Dust mites do not transmit any disease-causing pathogens.

5.6 Allergenic Properties of House Dust Mites

In mid 1900's, several scientists reported house dust mites as a major cause of the allergic properties of house dust (Spieksma and Voorhorst, 1969; Ranganath and Chennabasa Vanna, 1988; Johansson and Van Hage-Hamsten, 1990; Revsbech and Ducholm, 1990). Later on, the allergenic significance of these mites and their inter-relationships has been studied in detail (Spieksma and Voorhorst, 1969; Johansson and Van Hage-Hamsten, 1990; Ranganath and Chennabasa Vanna, 1988; Revsbech and Ducholm, 1990). The first reports of asthmatic symptoms produced by these mites appeared in mid 1900's (Ancons, 1923; Storm *et al.*, 1924). Mites cause dermatitis, other skin infections (Schamber, 1910; Hirst, 1912; Swan, 1934) and respiratory allergy (Young *et al.*, 1990).

Mite allergy caused by house dust mites is becoming increasing hygienic concern in many countries of the world. According to some, they are playing a useful role in a biological/ ecological chain in the man's anthropogenous niche (Cohen, 1928; Wharton, 1976; Van Bronswijk, 1978; Burr *et al.*, 1980; Gebrial *et al.*, 1982; Spieksma, 1990), as their waste products (excreta) are capable of sensitizing humans (Fain *et al.*, 1988; Platts-Mills and DeWeck, 1989).

Presence of dogs is positively correlated with allergen levels. Carpets, beds, upholstered furniture are considered as primary source of allergens. Antigen levels are higher during hot and humid season. Higher antigens levels are found in bedroom carpets, wooden structures (Murray and Ferguson, 1983; Konishi and Uehara, 1994; 1995) and concrete structures (Uchikoshi *et al.*, 1982; Osaka *et al.*, 1985).

5.7 Mechanism of Mite Allergic Reactions

The humoral immune response to mite antigens is similar to that of other antigens (Platts Mills and Chapman, 1987). The immunoglobin (antibody) IgE is produced following the chain reaction given below, is the mediator of allergic reactions.

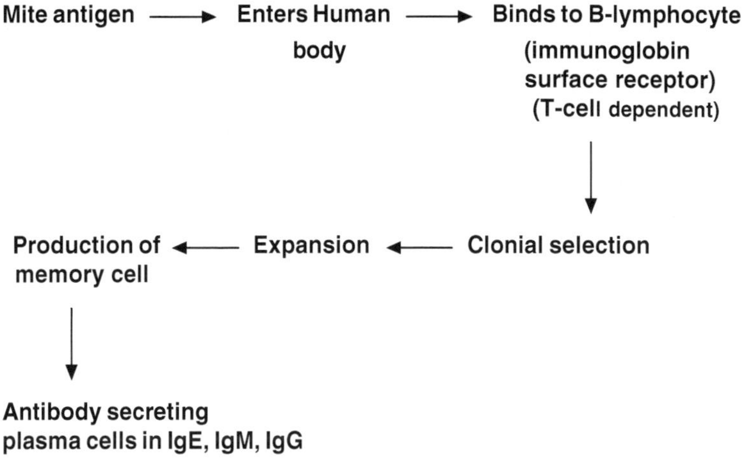

IgE is present in very small amounts in the serum of normal individuals (0.004 per cent of total serum immunoglobin or 0-0.002 mg/ml). In individuals predisposed to allergy (genetically), plasma cells produce IgE in response to their first introduction to an allergen from mites. The IgE binds to the surface of mast cells (concentrated in the respiratory and gastrointestinal tracts and skin) and basophils (blood). Upon subsequent encounters with allergenic molecules, interaction between allergen and IgE triggers the release of chemical mediators such as histamines, prostagelandins and lenkotrines from the mast cell. These mediators cause itching, vasodilation, construction of smooth muscle, changes in membrane permeability and leakiness of blood vessels. The clinical signs of allergy include runny and itchy nose, sneezing, tearing, red and itchy eyes, eczema, asthma, abdominal cramps and diarrhea.

5.8 Diagnostic and Immune Assays

Because of the central role of IgE antibodies in the allergic reactions, most of the diagnostic and immune assays directly or indirectly measure bound or serum levels of IgE. Skin tests measure an allergens interaction with IgE on mast cells in the skin. Radioallergosorbent test (RAST), enzyme linked immunosorbent assay (ELISA), crossed radioimmuno-electrophoresis (CRIE) measure circulating IgE levels in human serum. Leukocyte histamine release

(LHR) assays, immunoblotting by SDS-PAGE (Sodium dodecyl sulfate-polyacrylamide gel electrophoresis), isoelectric focusing (IEF), single and double immunodiffusion, rocket immunoelectrophoresis and crossed immuno electrophoresis (CIE) are some of the other diagnostic methods to measure IgE. IgE RAST assay using paper discs coated with *D.p., D.f.* or *E.m.* extracts are available for immunodiagnostics to identify source of allergens (Konishi and Uehara, 1990). The International Union of Immunological Societies (IUIS) has given a system for nomenclaturing the mite allergens. Accordingly, the first three letters of the genus (italicized); space; first letter of species name (italicized); space; a Roman numeral to designate the chronological order in which the allergen was isolated, are given for the isolated and characterized mite allergens. Till date, *Der p.* I, *Der p.* II (from *Dermatophagoides pteronyssinus*) (Chua *et al.*, 1990), *Der f.* I, *Der f.* II (Heymann *et al.*, 1989), *Der f.* III (from *D. farinae*) and *Der M.* I (from *D. microceras*) are characterized.

5.9 Important House Dust Mites

5.9.1 *Dermatophagoides pteronyssinus* (Trouessart)

It is commonly known as the European house dust mite (Figure 5.2). Both male and female adult house dust mites are globular in shape, creamy white and have a striated cuticle. The female is measured approximately 420 µ in length and 320 µ in width. The male is approximately 420 µ long and 245 µ wide. A pair of suckers on the ventral posterior idiosoma of the male is used to grasp the female during copulation. Males are more sclerotized with enlarged

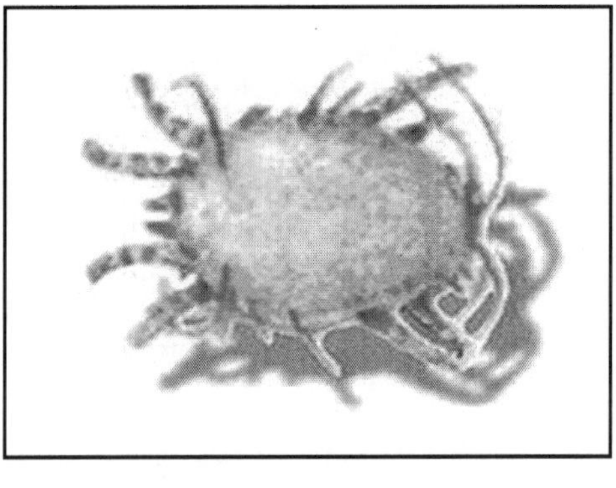

Figure 5.2: ***Dermatophagoides*** **sp.**

legs I and III. The male aedeagus is located between the apodemes of leg IV. Mated females live about two months. A male may attach itself to a tritonymph female and mate when she reaches the adult stage. It lays about 80 eggs over a 45-day period. These mites feed on shed skin of man. The average individual sheds 0.5 to 1.0 g of skin daily.

Egg hatching period, larval, protonymphal and tritonymphal periods are 4.48, 6.39, 4.39 and 4.03 days, respectively at 30°C (Arlian, 1992). Females per day lay four eggs and eggs are laid singly. Quiescent period is about two days after each preceeding active period, but it increased to almost double if temperature drops down to 23°C (Arlian *et al.*, 1990). Under unfavourable conditions, diapause of quiescent protonymph occurs and may last for months before tritonymph emergence. Arlian *et al.* (1990) observed 123 days for quiescent protonymphs at 15°C, which showed significantly reduced metabolic rate. These are observed as firmly glued to the substrates in cultures although under optimal conditions, this stage did not form (Arlian *et al.*, 1990). At 20-30°C, the life cycle is completed in 18-35 days.

5.9.2 *Dermatophagoides farinae* Hughes

It is commonly known as the American house dust mite. *D. farinae* lays eggs over a 30-day period producing about an egg a day. One generation from egg to egg is completed in 18-35 days at 20-30°C (Furumizo, 1975; Anderson, 1988; Hart and Fain, 1988). Like *D. pteronyssimus*, *D. farinae* also went in quiescent protonymph stage under below optimal conditions. It lasts for 143.8 days at 15.6°C (Ellingsen, 1974; 1978; Arlian *et al.*, 1983).

5.9.3 *Euroglyphus maynei*

This mite is also considered as an important agent in causing allergenicity (Voorhorst *et al.*, 1969; Colloff, 1991). Pre-oviposition period is three to four days (Taylor, 1975) although Hart and Fain (1988) reported this period to be fourteen days. Eggs are laid in small groups of 3-4 eggs. Larva, protonymph and tritonymph period is about 10-17, 5-17 and 6-12 days, respectively (Nannelli *et al.*, 1983). At 25°C and 75-80 per cent relative humidity, development time from egg to adult is recorded 30-35 and 50-53 days for male and female mites, respectively.

Higher mean density of *E. maynei* in mattresses (136/100g dust) is recorded by Blythe *et al.* (1974) than in bedroom floors, living rooms or kitchen. The mite is recorded from all over the world with maximum population in Kenya (158.76/100g; Blythe, 1976), Venezuda (122/100g; Hurtado and Parini, 1987), UK (129/100g; Blythe *et al.*, 1974), India (100.9/100g; Ranganath and Channabasavanna, 1988) and Germany (99.9/100g; Keil, 1983). Hatching period (1-24h) is followed by an active non-feeding period in which larvae become swollen and sessile.

5.9.4 *Glycyphagus domesticus*

These small mites (Figure 5.3) are very similar to those found in foodstuffs but under a lens they can be recognised by the long hairs at their rear end. They may occasionally occur in foodstuffs, but they are primarily found in old-fashioned upholstered furniture, which has been kept in a damp area for some time so that the stuffing has rotted. House mites then feed on the fungus and multiply in large numbers. They occur particularly in upholstery stuffed with fibres derived from palm leaves. Now days these mites don't present a problem with modern furnishings and central heating because like all other mites they quickly die if they become desiccated.

5.10 Contamination

Contamination through fungal spores is noticed in all the three

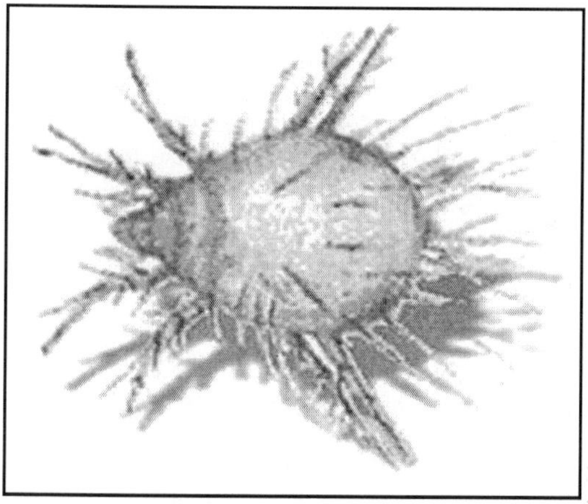

Figure 5.3: *Glycyphagus* sp.

species. Nine species are isolated from *D. pteronyssinus,* out of these *Aspergillus, Eurotium, Wallemia* species, are of common occurrence (Hay *et al.,* 1992). Low relative humidity is found to be an important factor for this contamination.

5.11 Control Measures

Allergies from house dust mite can be managed by immunotherapy using mite extracts in conjunction with reducing the mite population in the home of the patient. It is virtually impossible to eliminate all dust mites from a household, but areas where mites tend to congregate can be targeted. Wall to wall carpet should be reduced and replaced where possible with smooth hard surfaces that are easy to clean. Upholstered furniture can be avoided, mattresses and pillows should be encased in especially manufactured ('plastic') products that aid in the exclusion of mites. Bedding and curtains should be selected on their ability to withstand frequent laundering. A reduction in humidity levels by increased air circulation and ventilation throughout the house will assist in maintaining mites at a lower level.

Physical control by heating or drying (Oshima *et al.,* 1972; Korsgaard, 1983) of infested material reduced the population significantly. Exposure to sun effectively control mite infestation as it kills all the stages of pyroglyphid mites due to drying. Regular cleaning, sweeping and maintaining hygienic conditions are also considered effective in checking allergic mite population. According to some, vaccum cleaning is also an effective method to remove the mites from beds, upholestory items, carpets and other textile objects. However, as house dust mites clings to solid supports, with vacuum cleaning living mites can only be removed from the area of a textile object where the air stream is sufficiently strong *i.e.* in the upper area of the textile structure (Bischoff, 1988). Normal vacuum cleaning (2 min./m^2 with upholestry nozzle) removes approximately ten per cent of the house dust mite population present in particular object. This may be the reason that Bischoff and Von Bronswijk (1986) did not considered it as an effective long term measure. Flat textiles are treated more effectively by vacuum cleaning than voluminous ones. In the same way, carpets with vertical fibres are cleaned more effectively.

These mites are sensitive to relative humidity and at 60 per cent or lower the mite population stops growing and dies out. A number of insecticides and fungicides are reported to provide some relief but no method has been found for reducing mites and relieving allergy suffering. Immunotherapy, *i.e.,* injections of mite extracts into the patients to increase antibody level, has had variable success (Munro-Ashman *et al.* 1976).

Acarines predators of cheyletidae family are of common occurrence in house dust samples especially *Cheyletus* species. But this species is not effective on pyroglyphid mites because the optimum humidity for *Chelytus* species is 90 per cent and pyroglyphid mites thrive below this humidity, sometimes even at 60 per cent relative humidity (Hart and Whitehead, 1990). Mites can be important source of airborne allergens, especially on farms. The increasing occurrence of allergies in the developed countries and greater interest in food safety has drawn attention to the presence of allergens in agro products during storage. Stored agro-products can contain allergenic contaminants of plant, microbial, or animal origin. Dust mites are the main cause of asthma, which thrive in dust of the houses. In one case study, conditions of 80 percent of asthmatic patients worsen due to dust mites. It is not visible to naked eyes but can only be seen through microscope. Dust mite grows rapidly in humid places (optimum 75 per cent).

Chapter 6

Parasitic Mites

Parasitic mites associated with insects, birds and mammals (rodents, small mammals, human beings) are known, out of which majority of them are ectoparasite and few are endoparasite especially in respiratory tract. These are primarily blood sucking mites but some may feed on the lymph or dermal scrappings. Mite infestation causes irritation, dermatitis (itching inflammation) and allergic reactions in host, which are termed as acarisis. Skin mites (*Sarcoptes scabei, Notoderas cuniculi, Psoroptes ovis, Cheyletiella parasitiverase*), fur mites (*Radfordia* sp.), hair follicle mites (*Demodex* sp.) cause direct damage to their host. Scratching and irritation caused by mite create a congenial environment for secondary infection (Boyd, 1951).

In animals, they affect lips, chin, nose, face, head, roots of ear and circles around eyelids. Animals show yellowish or greyish crust, lesion on the body and drooping of ears (Zumpt, 1961; Fain, 1965; Alwar, 1970; Prasad, 1974; Putatunda, 1979).

Parasitic mites display an amazing range of structural and biological adaptations to suit their microhabitats like skin, hair, feather or respiratory tract of animals. The most efficient among them adjusted, thrived, multiplied and became successful as parasites. An unusual phenomenon, Neosomy occurs in the life cycle of parasitic mites associated with insects (*Pyemotes* sp.–ectoparasite of stored insects), reptiles (*Vatacarus* sp.–in snake lungs) and mammals (*Riedelinia* sp.–on bat wings) in which larvae or adults enlarged to 100–1500 fold and develop a papilliform neosomule which facilitates attachment to host or transport through the mucous of the host.

The parasitic mites parasitize not only on invertebrates and vertebrates including man but also play an important role in the transmission of various diseases. Gamasid mites are considered to be vectors of epidemic haemorrhagic fever virus. In China, *Haemolaelaps glasyonoi, Eulaelaps stabularis* are reported as vectors of this virus (Table 6.1). *Leptotrombidium akamushi* and *L. deliense* are proved transmitters of scrub typhus in man. Laelapid mites transmit tularaemia among rodents.

6.1 Mites Associated with Mammals

Small mammals such as rats, squirrels, bats, rabbits and guinea pigs live in close association of human beings and harbour large number of mites which have been studied in detail (Yunker, 1973; Ogata *et al.*, 1978; Wikerhavser *et al.*, 1986). Mostly, these belong to three suborders–Astigmata, Mesostigmata and Prostigmata. These parasitic mites are habitat specific rather than host specific. With in host body, distribution of different species depends on the less chances of attack on them, which increased their probability of survival. Chigger mite is primarily found on host head whereas *Gahrliepia* is encountered in the ears of house rats away from direct predation. Among the Laelapidae, *Gigantolaelaps oudemansi* Fonseca, *Androlaelaps* (*Haemolaelaps*) sp., *Gigantolaelaps gilmorei* Fonseca; are the most frequent species associated with rats (Table 6.1). No significant relationship is found between parasite intensity and host sex for any host species. The relationship between parasite intensity and period of the year (dry or wet) showed higher values in the wet (1.95) than in the dry (1.72) period. Thoracic region and base of tail are the main regions infected by mites on rats.

6.1.1 Biology

Mite associated with mammals and birds are usually translucent or whitish in colour so they are difficult to detect. After feeding on host's blood their colour changes to reddish brown. Life cycle includes egg, larva (6 legs), two nymphal stages and adult. It is difficult to study the biology of parasitic mites, which penetrate the host skin, and lays eggs there. Wherever information is available for particular mite, it is incorporated under relevant heading.

Table 6.1: Parasitic Mites and Associated Disease Symptoms in Mammal

Mites	*Hosts*	*Disease/Symptoms*
Apodemus sp.	Small mammals	Haemorrhagic fever virus
Chirolaelaps mystacina	Bats	Heavy infestation kills host
Eulaelaps stabularis	Human beings	Haemorrhagic fever virus
Haemolaelaps glasyonoi	Human beings	Haemorrhagic fever virus
Laelaps echidninus	Rodents	Tularemia
Leptotrombidium akamushi	Small mammals	Scrub virus
L. deliense	Small mammals	Scrub virus
Liponyssoides sanguineus	House mouse and other rodents	Rickettsial pox
Mycoptes musculinus	House mouse	Increased mitotic activity, foot pad lesions
Neotrombicula autumnalis	Human beings	Itchy red spots
Ornithonyssus bacoti	Animal technicians	Dermatitis
P. ovis	Small mammals	Severe skin inflammation
Psoroptes cuniculi	Small mammals	Severe skin inflammation
Sarcoptes scabei	Small mammals	Scabies

6.1.2 Nature of Damage

These mites act as causative agents for anaemia, pneumonia, crusting of skin, hair loss, decreased reproduction and death of host occurs in severe case of infestation. Inflammation, scratching and irritation lead to secondary infection (Figure 6.1). *S. scabei* form tunnels into the skin and in case of heavy infestation may even cause death of the animal. *P. ovis* cause inflammation in rabbits. Haemoglobin is derived from small haemorrhages on skin, which cause severe inflammation due to deposition of antigenic material. In *P. cuniculi* infestation, disease is symptomised by shaking of head, scratching of ears and crust formation on the pinna. Heavy infestation may cause fatal meningitis. *Chirolaelaps mystacina* infestation in bats cause severe irritation that leads to self inflicted injuries and provide congenial place for secondary infection by microorganisms. Listophorid mites are attached to the mammalian hairs and feed on the secretions of sebaceous glands.

Antigens of mange mites, *P. cuniculi* and *P. ovis* showed cross reactivity with the antigens of house dust mite, *D. pteronyssinus*. Likewise *S. scabei* is cross reactive with both *D. pteronyssinus* and *D. farinae*.

6.1.3 Important Mites Associated with Mammals

Several species of mites associated with mammals may be found in homes (Figure 6.2). These include the house mouse mite, the rodent mite and the rat mites. All of these mites are associated with wild or domestic rodent populations. The life cycle of these mites generally requires 17 to 23 days. If these mites are found, they almost certainly indicate the presence of a rodent infestation in or around the building.

6.1.3.1 *Sarcoptes scabei* (De Geer)

This mite is commonly known as Itch or Scabies Mite. This mite is very seldom seen, lives actually in and on the skin, with the female burrowing tunnels in the upper horny layer, particularly on the front of the wrists, the sides and webs of the fingers, the buttocks, the genitals and the feet. Different varieties of *S. scabei* are believed to be specific for different mammals, including man and a large variety of domestic and wild animals, but are transferable from one host to another. Itch mites tunnel into the skin, especially on hands and wrists. Nodules burst and ooze serum, later hardening to form scabs. The skin between the fingers is often invaded. Itching is intense and known as «scabies» or «seven-year itch.» Scratching can cause bleeding and infection of open sores. Overlapping generations occur at two to three week intervals.

They are so small that even the adults are barely visible to the unaided eye. Adult females are 0.33 to 0.45 mm long, and the males, 0.20 to 0.24 mm. The mites are translucent, dirty-white in colour. The more highly chitinized portions appear as brownish. In living specimens, the body is divided into two regions by a fold in the integument (any natural protective covering, such as a skin, rind, or shell); the posterior portion bears the last two pairs of the four pairs of very short legs. The last two pairs of legs do not extend as far as the margins of the body.

Life Cycle

Both sexes and all stages of the itch mite tend to burrow into the skin immediately when placed on it. The nymphs and males make only small, temporary holes, whereas, the egg-laying female made the largest and longest burrows. The female always burrows in folds of the skin, preferring the deeper furrows and cracks. She can be induced to enter when a fine scratch has been made with a needle in the surface of the skin. She may also place herself in the acute angle between a sloping hair and the surface of the skin to gain support for initiating the burrowing. The winding burrow may reach a length of 5 to 15 mm.

The adult longevity is from 2 to 6 weeks. Although it is difficult to determine the number of eggs lay by a female in her lifetime, but it is usually estimated to be between 40 and 50. The period between the beginning of burrow formation and the finding of the first larva varied from 71 to 78 hours. The duration of the egg, larva, protonymph and deutonymph is 2.5–3.5, 1.5 –3, 1.5–2.5 and 2 –4 days, respectively. The life cycle is completed in 9-15 days. The parts of the body selected by the ovigerous female are the interdigital spaces, the wrists and the ulnar margins of the wrists, the elbows and the anterior folds of the axillae, the penis, scrotum, buttocks, the back of the knee, the ankles and toes. In young children, the egg burrows may occur on any part of the body, and in women, the undersides of the breasts are very commonly selected.

Means of Transmission

Transmission is usually by direct contact with an infested person or through infested clothing and bedding. People are most likely to become infested when living in continually crowded quarters, such as slums or jails, or during periods of major calamities that result in prolonged overcrowding.

Symptoms of Infestation

In animals large numbers of mites give rise to "sarcoptic mange," whereas in humans relatively small numbers of mites can cause unpleasant symptoms, and the disease is known as "scabies." At a certain stage, the irritation may become so severe that the patient becomes frantic and suffers from lack of sleep. If the infestation is long continued, or if a later infestation occurs, an allergic reaction develops, with intense itching and a redness, or rash of follicular papules (small blister at the base of a hair) over major part of the body. Head, ears, legs are preferred by mites for initial infestation then they become generalized in the whole body. The rash may develop on areas such as around the armpits, the wrists, the waist, inside the thighs, and backs of the calves.

Low temperature and high relative humidity increased its survival (Arlian *et al.*, 1989). Females and nymphs are recorded as more resistant to desiccation than males and larvae. Dislodged mites are found to survive up to several days to several weeks in absence of hosts. This quality indicated their importance in scabies transmission and reinfestation.

6.1.3.2 *Psoroptes cuniculi*

This is basically found to be associated with sheep and goats. In Brazil, prevalence of *P. cuniculi* was recorded as 46-83 per cent in sheep and 78-89 per cent in goats (Faccini and Costa, 1992). Occurrence of Otitis in goats and psoroptic Otocariasis in sheep (Shastri and Deshpande, 1983) are also reported. Life cycle includes egg, larva, protonymph, tritonymph and adult.

6.1.3.3 *Ornithonyssus bacoti* (Hirst)

This mite is commonly known as Tropical Rat Mite. This mite will feed on humans even when rats are present, causing painful bites, intense itching and dermatitis. Barely visible, they are bright red to black with white markings and are usually noticed on walls in basements, kitchens, bathrooms and where rats are found. Mites drop from their host after each blood meal and can survive several days without feeding.

6.1.3.4 *Liponyssoides sanguineus* (Hirst)

This mite is commonly known as House Mouse Mite. This mite is found on mice and can bite humans. It will wander away from mice on to walls.

6.1.3.5 *Demodex folliculorum* (Simon)

This mite is commonly known as Follicle Mite. These microscopic mites live in the hair follicles or sebaceous glands of most humans. Very few persons are allergic to them but those who are, may lose their eyelashes or develop acne.

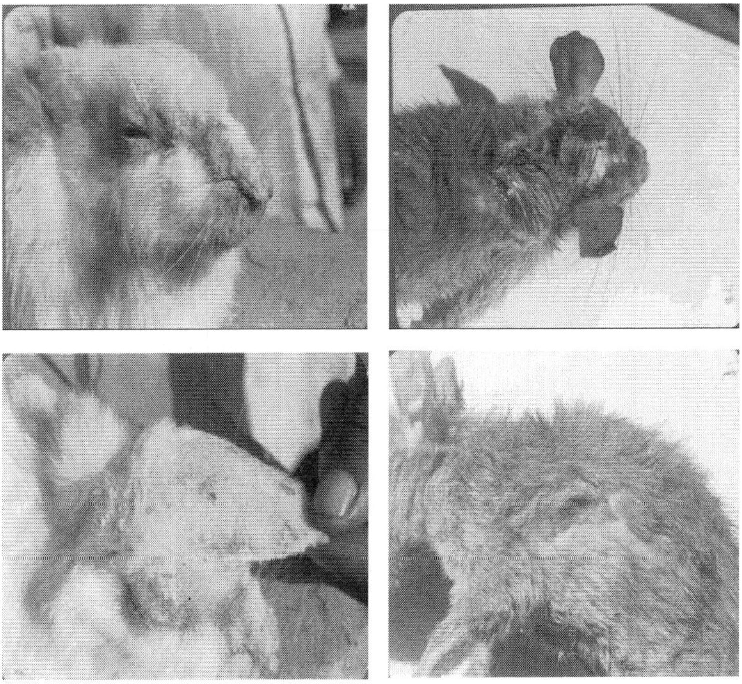

Figure 6.1: Damage Caused by Parasitic Mites (Rathi, 1991)

6.1.3.6 *Pyemotes tritici*

This mite is commonly known as Straw Itch Mite. These mites can cause epidemics of dermatitis during harvesting and post-harvesting operations in straw, hay or certain grains. Bites produce a rash-like dermatitis, extending over much of the body and accompanied by itching, sweating, fever, headache and even vomiting in severe cases. They are beneficial because they feed on larvae of wheat jointworm, rice and granary weevils, Angoumous grain moths and other pests.

6.1.4. Control Measures

Before initiating treatment, it is important to diagnose the disease correctly, for neither the irritation nor the liability to skin diseases can cease until the mites have been eliminated. In case of heavy irritation, doctor should be consulted immediately for necessary

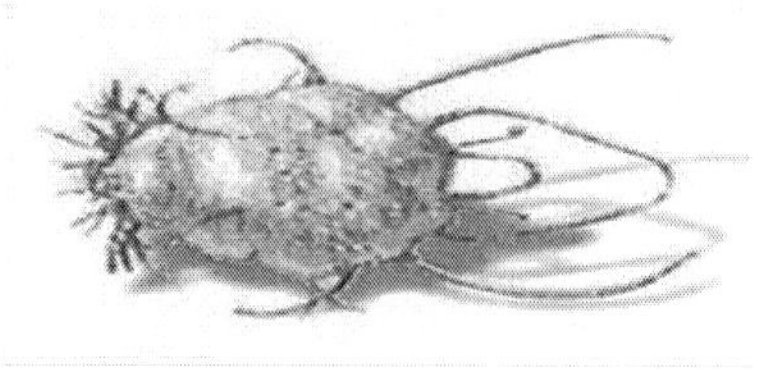

Sarcoptes **sp.**

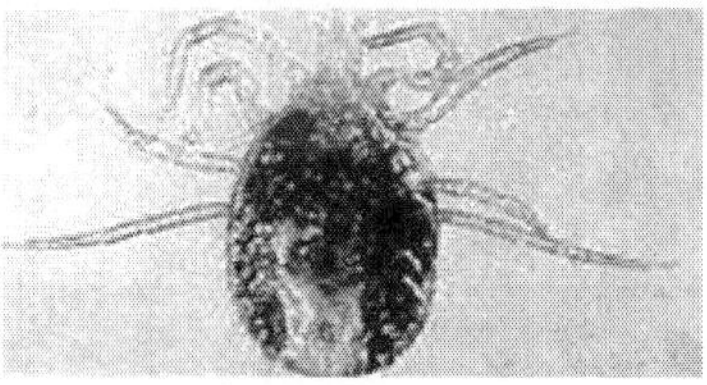

Ornithonysus **sp.**

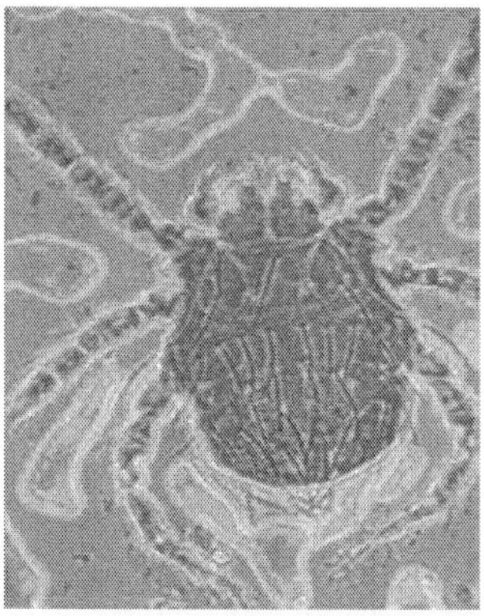

Trombicula **sp.**

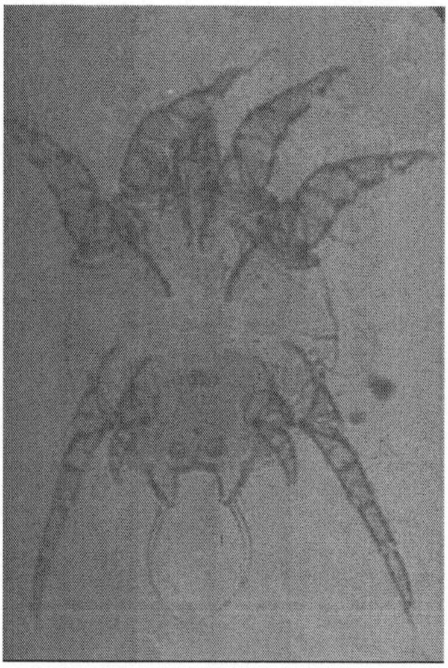

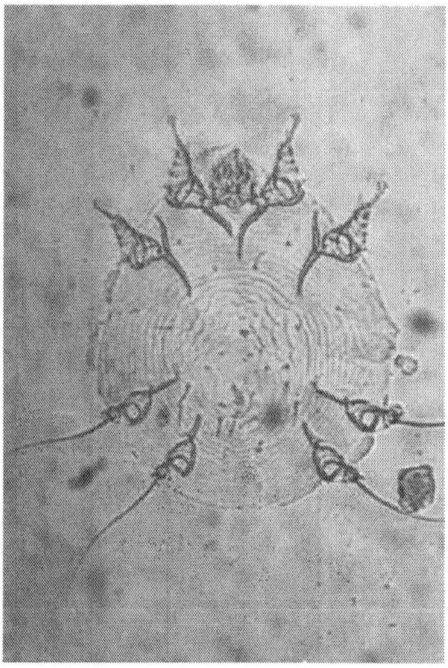

Figure 6.2: Parasitic Mites Associated with Birds and Mammals

treatment. A warm bath with medicated soap is also helpful in reducing the skin mite infection in animals and humans. In case of human beings, putting clothing away for 2 or 3 days at ordinary room temperature should be sufficient to get rid of mites, as they dessicate very easily. Other precautions like removal of mammal nests etc. are common practices observed for bird mite infestation, which are discussed in bird mites.

6.2 Bird Mites

These mites are parasites on birds and are usually found on the birds or in their nests. Thus, under normal circumstances, bird mites go unnoticed. Occasionly, bird mites wander away from the bird nests and are found indoors as accidental invaders.

6.2.1 Distribution and Bioecology

Migration by bird mites is common in late spring when the baby birds are leaving the nest. Mites that migrate away from the nests may be found anywhere in houses and buildings, but are most abundant in the vicinity of nests that are built on or in the house. Sparrows, starlings and pigeons are the birds most often associated with bird mite infestations in buildings. The chicken mite and the northern fowl mite are both ectoparasites of poultry and other domestic fowl, as well as many species of wild birds. The life cycle is very short (7 days), and if conditions are favourable, they can rapidly generate large populations. The bites of these mites cause itching, swelling and raised reddish spots on the skin. Bird mites are semi-transparent in colour, which makes them difficult to detect on skin until blood is ingested and then digested; when they may appear reddish to blackish. The mites feed on the unfeathered nestlings, as well as the adult birds, and the large amount of nesting material used by the birds provide the mites with an ideal environment to thrive.

Ornithonyssus bursa is one of the common mites associated with birds, which invade human homes. Apart from this, several other mites can invade dwellings and bite humans. These bird mites include *O. sylviarum* (Northern fowl mite) and *Dermanyssus gallinae* (Chicken mite). Also, a closely related species, *O. bacoti* (Tropical rat mite), occasionally attacks humans. This species is associated with rodents, such as rats and mice, and their nests and discussed under mites associated with mammals.

Bird mites are either commensals feeding on debris and oil secretions on feather or parasites living inside quills or in the skin. Among the parasitic bird mites, feather mites are reported to cause economic damage and generally encountered along the shaft of feather and in more intense infestations, run along the barbs and barbules. Incidence of feather mite depends on the host specificity (Cerny, 1973; Atyco and Perez, 1990), site selection among different regions of the body (Dubinin, 1951; Choe and Kim, 1988), feather types (Peterson, 1975; Perez and Atyeo, 1984) and regions of the feather (Atyeo and Perez, 1988; Choe and Kim, 1988). The population dynamics of these parasites, seasonal changes in infestation rates (McClure, 1989), their ecology in relation to other ecto-parasites (Choe and Kim, 1987), difference in prevalence of infestation between host species (Wheeler and Threlfall, 1986), host ecology and social behaviour (Poulin, 1991; Poiani, 1992) have also been documented. Behnke *et al.* (1995) categorized birds into four groups depending on their infestation intensity with mites.

6.2.2 Infestation Level in Birds

Highest and intense infestation: Swallows, Sand martins, Green finches.

Moderate level infestation: House sparrows, Black birds, Serins, Black caps, Gold finches, Warblers, Great tits.

Low level infestation: Nightingales, Short toed tree creepers.

Non detectable mite infestation: Wrens, Fan-tailed warblers, Waxbills.

Mite infestation in birds leads to deplumation, mange, tumor, scaly legs, lameness, pneumonia and anaemia and in severe infestation cases, even death of host is reported (Boyd, 1951; Zumpt, 1961). Avian mites often transmit cholera producing bacterium, *Pastuerella*, bird protozoa *Hepatozoon* and *Toxoplasma* (Boyd, 1951). In India, 96 species of bird mites are recorded which are associated with 60 bird species as host (Putatunda *et al.*, 1981; 1989; Singh and Putatunda, 1999).

6.2.2.1 *Trombicula autumnalis*

It is commonly known as Harvest Mite. It is not the adult stage of the Harvest mite, which causes the damage, but the six-legged larval stage, which causes continuous scratching and leads on to secondary infection. These are white to orange red in colour and very small (1/100 of an inch) in size.

Adult harvest mite overwinter near or slightly below the soil and in other protected places. Females become active in the spring and lay up to 15 eggs per day in vegetation when soil temperatures are 60°F. The larvae congregate in groups on small clods of earth, in matted vegetation and even on low bushes and plants, where they have more access to a prospective host. The first active stage in the life of the harvest mite is the six-legged larva and this is the only stage, which attack warm-blooded animals.

6.2.2.2 *Ornithonyssus bursa*

Bird mites" or "Tropical fowl mites" are the common names used to describe the mite *O. bursa* from Macronyssidae family. It is widely distributed throughout warmer regions of the world. The mites are haematophagous (feed on blood) and natural parasites of common birds including pigeons, starlings, sparrows, Indian mynahs, poultry and some wild birds. The mites are small with eight legs, barely visible to the eye, oval in shape with a sparse covering of short hairs and are extremely mobile. They are semitransparent in colour, which makes them difficult to detect on skin until blood is ingested and then digested. After this, they may appear reddish to blackish.

Contact with humans occurs after the birds gain entry through unprotected areas, or to roof cavities via broken tiles to construct their nests in homes, factories, barns and other dwellings in early spring or summer. The large amounts of nesting material used by the birds provide the mites with an ideal environment to thrive. The unfeathered nestlings and the adult birds occupying the nest are utilised as a ready food (blood) source for the expanding population of mites. When the young birds are ready to fledge, some mites remain attached to their feathers and feed intermittently. The mites remaining in the abandoned nest, and left without a suitable host when the birds leave or die, will roam and disperse throughout the dwelling over a 1-2 week period searching for new hosts. Most mites will die within 10 days without a blood meal from a bird host. They will bite but cannot survive on humans and do not infect/infest human skin.

Although the effect of a bird mite invasion is temporary, the mites can cause severe irritation, rashes and intense itching from the saliva they inject while biting. Scratching of the bites may lead to secondary infections. Also, the sensation of crawling mites on the skin can irritate some people.

Bird mites do not live underneath the skin and are not associated with any human diseases. The mites will bite at random and there are no selected areas of the body that are favoured, but clothed areas of the torso, limbs and the head area are commonly involved. The occupants of the building experience a great deal of discomfort until the infestation is controlled or dies. A bird mite infestation cannot be maintained on humans but will persist while the bird-related source of the mites is maintained.

6.2.2.3 *Hypodectes propus*

It is an inhabitant of the nests of columbiforus birds (Fain, 1967). The hypopus of this species lives in the subcutaneous tissues of the pigeon. The larval and protonymphal stages are reduced to apoderma and surround the developing deutomymph within the shell of the egg. On emergence, the small deutonymphs penetrate the skin of the nestlings and greatly increase in size. When host begins to incubate its eggs, these deutonymphs (hypopodes) pass through the skin into the nest. They then undergo rapid development into the adult phase. The female is non-feeding and produces numerous eggs, which develop into hypopodes in the nest.

6.2.2.4 *Dermanyssus gallinae* (DeGeer)

It is an ectoparasitic mite, commonly known as Chicken Mite, which causes economic damage in commercial poultry flocks (Axtell and Arends, 1990). It is found to feed on the blood of fowl, pigeons, cage birds, wild birds and few mammals including humans. This mite feeds on the host at night and hides in cracks and crevices during the day. It moves into crevices during day to digest blood meal and lay eggs (McGarry and Trees, 1991).

A new generation may occur every 5 to 7 days. Bites cause itching and light dermatitis. These mites enter a house after death of the host bird or departure from the nest. They can live for a month or more without a blood meal. Trap perches are found to be an attractive refuge for fed mites.

On the basis of LC_{90} values, carbamates were recorded as most toxic followed by organophosphates and pyrethroids.

6.2.2.5 *Ornithonyssus sylviarum* (C.and F.)

This mite is commonly known as Northern Fowl Mite, which causes economic damage in poultry flocks (Axtell and Arends, 1990). This barely visible mite completes development on the host bird (*i.e.*, poultry, pigeons and starlings) and leave the body of a dead host in large numbers to wander over walls, ceilings and bedding, seeking a new host. Bites can cause itching and dermatitis. This mite can survive two to three weeks away from the host.

Bendocarb, tetrachlorvinphos, primiphos methyl, permethrin, amitraz and carbaryl are observed as highly toxic whereas dichlorvos, malathion and fenvalerate are found to be of low toxicity.

6.2.3 Control Measures

Effective control of bird and mammal mites is achieved by following steps:

1. Locating and removing bird or rodent nests is the first step in bird mite control.
2. Treating infested areas (in the area of the bird nest as well as in the vicinity of the invaders) with an appropriate acaricide.
3. Additional spot treatments in areas where mites are seen wandering may be helpful.
4. Spray the acaricides on cracks, crevices and corners but not on counters, cupboards and other such general areas.
5. Initiate any structural changes or sanitary measures needed to prevent re-infestation by birds or mammals. Broken tiles or timber allowing access to roof cavities should be repaired and all potential entry points should be blocked.
6. Roosting and nesting sites on window ledges should be cleared and made unsuitable for future bird use.

6.3 Honeybee Mites

"The rapidity of the spread of *Varroa jacobsoni* over many parts of world demonstrates the mobility of modern beekeeping". The words expressed by Ruttner (1983) is also true in the present scenario with other associated parasites and pathogens of honeybees which are accidently spreading to new and uninfested areas with the increasing introduction of bee colonies to unexplored diverse fauna. Amongst these, more than hundred mites are associated with honeybees (Putatunda, 2001), which cause a mere nuisance to severe loss in colonies. Mites found in beehives, which are phoretic, or scavenger in nature has little or no effect on the bees. Parasitic mites, which feed and reproduce on bee brood and adult bees (Table 6.2), are serious threat to beekeeping industry. According to the nature of infestation, these mites are grouped into four categories.

Table 6.2: Mites Associated with Honeybees (*Apis* spp.)

Mite	*Hosts (Apis)*
Acarapis woodi	*A. mellifera, A. cerana, A florea*
A. dorsalis	*A. mellifera*
A. externus	*A. mellifera*
Acaropsis sollers	*A. mellifera, A. dorsata*
Varroa jacobsoni	*A. mellifera, A. cerana*
V. underwoodi	*A. cerana*
Euvarroa sinhai	*A. florea, A. mellifera*
E. wongsirii	*A. andreniformis*
Tropilaelaps clareae	*A. mellifera, A. dorsata, A. laboriosa, A. cerana, A. florea*
T. stimulis	*A. dorsata*
T. koenigerum	*A. dorsata, A. laboriosa, A. mellifera, A. florea*
Pyemotes herfsi	*A. cerana*
Neocypholaelaps indica	*A. cerana*
N. apicola	*A. mellifera*
Allodinychus flagellifer	*A. dorsata*

6.3.1. Ectoparasitic Mites

These mites are found attached to external surface of bees and feed on bee brood. These mites are considered as most serious which are threatening the survival of *Apis* species in many regions of the world. These mites act as external parasites of bees. They feed on egg, larva, pupa or adult. Some of the important ectoparasitic mites associated with honeybees are *Varroa jacobsoni, V. underwoodi, Euvarroa sinhai, Tropilaelaps clareae, T. koenigerum.*

6.3.1.1 *Varroa jacobsoni* (Oudemans)

This is one of the most serious and dreadful pest of bees (Matheson, 1993) in the world which led to the meeting of the European communities expert's group in 1983 for meaningful cooperation among the members to chalk out future strategies to provide a befitting reply to varroasis problem.

Initially discovered in Russia in 1964, it showed wide and rapid spreading from tropical to temperate zones and from *Apis cerana* (original host) to *A. mellifera* (adopted host). Grobov (1976) provided the figures on the rapidity of the spread of the mite over short distance. The mites from infested colony at a particular place were found in colonies 6-7 km away from initial host colony after three months. During the period, Grobov (1976) observed loss of more than two million honeybee colonies still creating havoc in many countries. In India, loss in pollen collection (57.2 per cent), brood rearing (46.2 per cent) and honey collection (66.9 per cent) are reported from mite infestation. Few lucky countries like Australia, New Zealand and Hawaii are reported to be free of this mite (Sammataro *el al.*, 2000). Mites are reported to transmit *Hafnia alvei* and *Serratia marcescens*, which causes Haffnosis and Septicemia, respectively, in bees (Glinski and Jorosz, 1992).

Morphology

It is comparatively a large (1.1 ´ 1.5 mm in size), dorsoventrally flattenned, reddish brown mesostigmatid mite exclusively found on *A. cerana* and *A. mellifera* and can be seen with naked eye. Parasitic adaptations include flattenned body shape, each tarsus modified into lobed sucker and presence of stiff hairs on ventral side, which entangles with hairy bees. It is commonly found between first two abdominal segments, head and thorax and thorax and abdomen. Mites easily penetrate intersegmental membranes of bees. From these regions, they feed on blood and cause wounds. Adult females are larger and darker than males. The female chelicerae are structually modified, the fixed digit is absent and movable digit is saw like blade capable of piercing and tearing the host integument.

Bioecology

V. jacobsoni is reported from *A. cerana* and *A. mellifera* (De Jong *et al.*, 1982; Schutz, 1984; Ifantidis, 1988). In *A. cerana* colonies, the damage is not considered serious because of its restricted reproduction only to drone sealed blood which comprises only 5-17 per cent of total brood area (Allen, 1965; Seeley and Morse, 1976) and its ability to remove mites through grooming behaviour (Peng *et al.*, 1987; De Jong, 1988; Tewarson *et al.*, 1992) in which they cut the parasitic mite from dentate mandibles.

In *A. mellifera*, due to reduced grooming behaviour of adults and mites ability to reproduce successfully in both worker and drone sealed brood had made them serious pest. The mites infest drone cells 7-8 times more than worker cells (Schutz, 1984; Fuchs, 1990) because of three reasons. Drone cells (45 L) remain attractive to mites for longer period of time prior to sealing than worker cells (15 h) (Ifantidis, 1988); the area of drone cell opening is 1.7 times larger than that of worker cells (Boot *et al.*, 1992) and drones receive more visits by bees due to their large size. Therefore, the role of mites breeding in drone cells could be important in enhancing the reproduction rate of mite at low infestation rates (Fuchs, 1986) during the initial build-up of the population.

The mite population increases from spring and is maximum in autumn upto three years, symptoms do not become noticeable. In fourth year, population is high (Fries *et al.*, 1991). When a number of mites infest the pupa of bee, the resultant adult has deformed legs and wings with short abdomen. The colony gets weakened and ultimately perishes.

Feeding on bee larvae blood is found necessary for egg laying by female mite (De Jong *et al.*, 1982). First oviposition occurred three days after brood cell capping and after every subsequent day another egg is laid with a maximum of five (De Jong *et al.*, 1982), six (Ifantidis, 1990; Martin, 1994) or eight eggs (Ifantidis and Rosenkranz, 1988). Offsprings developed and mate in bee brood cell. After reaching adulthood, the outer covering of young female hardens gradually (Langhe and Natzkii, 1977). Mother mites and viable offspring (adult females with a darkly pigmented dorsal shield) leave the capped brood cells with the emerging bee and begin phoretic phase attached to the bees and feeds on their haemolymph (Ritter, 1981; De Jong *et al.*, 1982). Lightly pigmented mites die soon after the young adult honeybee emerge from its brood cell. These non-viable females are not capable of surviving outside the brood cells, die after honeybee emergence and accumulate in the debris (Calatayud and Verdu, 1994). Females with hardened cuticle survive as phoretic mites on bees before entering into brood cell to reproduce. Eventually, darkly pigmented females drop on to hive bottom. Mortality of *Varroa* offsprings is much lower in drone cells than in worker cells (Martin, 1995). Duration of developmental time is recorded by various workers, which range between 6-10 days (De Jong *et al.*, 1982), 6-7 days (Martin, 1995) and 12 days (Fries *et al.*, 1994). Normal darkly pigmented mites which contribute 55 per cent of the total population produce four female adult offsprings /female (Martin, 1995). The longevity of female mites is recorded as two months in summer and 5-8 months in winter (Garg and Kashyap, 1998).

Symptoms

In the first three years of infestation, since population remains low, symptoms do not become noticeable. In the fourth year of infestation, presence of adult mites on bee body, spotty brood pattern, dead larvae, pupae, malformed or crippled workers and drones at hive entrance with shortened abdomen, deformed legs and wings, white droppings on empty cells and dead larvae with characteristic raised head (Garg and Kashyap, 1998) lead to decline and ultimate collapse of colony.

Control

Varroa mites can be treated with commercially available miticides. Miticides must be applied strictly according to the label in order to minimize the risk of contamination of honey that might be consumed by humans. Proper use of miticides will also help to slow the development of resistance among the mites.

Varroa mites can also be controlled through non-chemical means. Most of these controls are intended to reduce the mite population to a managable level, not to eliminate the mites completely.

Many beekeepers use a screened bottom board on their hives. When mites occasionally fall off from a bee, they must climb back up to parasitize a new bee. If the beehive has a screened floor with mesh of right size, the mite will fall through and cannot return to the beehive. The screened bottom board is also being credited with increased circulation of air, which reduces condensation in a hive during the winter.

Powdered sugar (Dowda Method), talc or other "safe" powders with a grain size between 5 and 15 micrometres can be sprinkled on the bees. The powder does not harm the bees, but does interfere with the mite's ability to maintain its hold on the bee. It is also believed to increase the bees' grooming behavior. This causes a certain percentage of mites to become dislodged. Powdered sugar works best as an amplifier of the effects of a screened bottom board.

Freezing drone brood takes advantage of *Varroa* mites' preference for longer living drone brood. In this case a frame is provided in the hive that is sized to encourage the queen to lay primarily drone brood. Once the brood is capped, the frame is removed and kept it in freezer. This kills the *Varroa* mites that are parasitizing those bees. It also kills the drone brood, but most hives produce an excess of drone bees so it is not generally considered a loss. After freezing, the frame can be returned to the hive. The nurse bees will clean out the dead brood (and dead mites) and the cycle continues.

Drone brood excision is a variation applicable to top bar hives. Honeybees tend to place comb suitable for drone brood along the bottom and outer margins of the comb. Cutting this off at a late stage of development ("purple eye stage") and discarding it reduces the mite load on the colony. It also allows for inspection and counting of *Varroa* on the brood.

Small cell foundation (about 0.3 mm smaller than standard) is believed to limit the space in each cell that *Varroa* mites have in which to inhabit and also to enhance the difference in size between worker and drone brood with the intention of making the drone comb traps more effective in trapping *Varroa* mites.

The Konya hive is also used to lower down the *Varroa* population. It has a cylindrical brood chamber, circular frames and an apparatus to rotate the frames according to a specific schedule. The rotation is believed to disrupt the *Varroa* mite reproduction cycle thereby reducing fecundity of the parasite.

Essential oils from mint family plants like peppermint, spearmint and lavender are used to control honeybee mites. The essential oil is mixed with light carrier oil such as canola or sunflower. A paper towel is saturated with this oil and placed it on top of the frames in the top super (the top box of the beehive). With time, the bees tear up the paper towel and remove it from the hive. In the process the mixture is distributed throughout the hive. The oil mixture is a mechanical mite deterrent as it inhibits the movement of the mite throughout the hive. The mixture is also a chemical deterrent as the mint oil is toxic to the mites, but not to the bees.

Several chemicals including fluvalunate and commaphos are used either as impregnated plastic strips (Borneck, 1985) or on plywood boards (Lubinevski *et al.*, 1988) in between the bee frames for *Varroa* control. However, resistance to organophosphates, fluvalinate and pyrethroids (Milani, 1995) are now reported worldwide.

Removal of drone cells, creating broodlessness, use of 30-85 per cent concentration of formic acid, essential oils and spraying dehydrated oxalic acid: Sugar: Water in the ratio of 1:10:10, biological control and resistant honeybee production are some of the useful control measures applied to *Varroa* control.

6.3.1.2 *Euvarroa sinhai*

It is found in Asia from Iran to Sri Lanka and is a parasite of *Apis florea*. It develops naturally on the capped drone brood (Mossadegh and Komeili, 1986). Development is completed within a week. Each female produces 4-5 offsprings. Workers and drones are used for dispersal (Sammataro *et al.*, 2000). In the colony infested by this mite there is hinderance in the construction of queen cell (Aggarwal and Kapil, 1988). Another species, *E. wongsirii* parasitizes the drone brood of *A. andreniformis* in Thailand and Malaysia.

Morphology

It is a large, brown, broadly pear shaped, setaceous mite which lacks fixed cheliceral digit (Delfinado and Baker, 1974). The male chelicerae are modified for sperm transfer and it is a non-feeding stage.

Biology

This mite attacks only drone brood. Males, nymphs and eggs are observed inside the sealed drone brood cells; however, adult females are also recorded from worker bees (Kapil and Agarwal, 1989). Population is found to attain peak in March-April and September.

6.3.1.3 *Tropilalaeps clareae*

This mite is a native pest of *A. dorsata* (Laigo and Morse, 1968; Delfinado-Baker *et al.* 1985) which is also reported on *A. mellifera*

(Sevilla, 1963; Stephen, 1968; Crane, 1968; Atwal, 1969; Atwal and Goyal, 1971; DeJong *et al.*, 1982; Nyein and Zmarlicki, 1982; Burgett *et al.*, 1983; Bhaskar and Putatunda, 1989; Matheson, 1993), *A. cerana* (Delfinado and Baker, 1982) and *A. florea* (Kapil and Agarwal, 1987). In India, it caused 50 per cent mortality in late larval and pupal stages of *A. mellifera*. Two peaks in *T. clarae* population (March-April and September-October) are recorded by Agarwal and Kapil (1989).

Morphology

Mites are oval or elliptical in shape with female comparatively larger (0.99 mm long) and darker (reddish brown) than male (0.91 mm long, light brown). Male chelicerae are toothed and used for sperm transfer. Gnathosoma is completely hidden by dorsal shield. The opisthogenital plate overlaps and plate in the female but is separate in the male. It has bidentate fixed upper digit and pointed movable digit, unlike *Varroa*.

Biology

Mating occurrs in brood cells with no pre-mating activity. It lasts for fifteen minutes (Rath *et al.*, 1991). The duration of egg, larval, protonymphal and deutonymphal stage is 1-1.2, 0.67-0.83, 2.5-2.8 and 2.9-3.1 days, respectively (Garg and Sharma, 1994). The total developmental period of female and male is 7-8 and 6-7 days, respectively.

Infestation in worker brood is recorded maximum followed by worker bees and drone brood (Garg and Sharma, 1994). Woyke (1985) recorded 2-54 per cent worker brood infestation by *T. clareae* and further reported 1.3-1.9 offsprings per female with 1.3-2.3 females per brood cell (Woyke, 1987).

Symptoms

Presence of brown fast running mites on comb, irregular and spotty brood, depressed sunken cappings of brood cells and malformed larvae, pupae and adult bees are the major symptoms associated with *T. clareae* infestation (Hosamani *et al.*, 2005; 2006). During heavy infestation, Worker bees with mutilated wings and stunted abdomen at the entrance, no egg laying by queen are also recorded which lead to broodless condition in colony.

Control

Both chemial and non chemical methods have been tried but work has not been extensive. Treatment with fluvalinate and formic acid (Garg *et al.*, 1984), dusting with sulphur (Sevilla, 1963) are found effective. Nyein and Zomerlicki (1982) recommended depriving *A. mellifera* colonies of brood by either removing brood frames or by caging queen for 21 days to control mite infestation. Thymol crystals (10g per colony) are also found effective against *T. clareae* (Garg and Sharma, 1988). Manipulation of brood rearing cycle in which mites are deprived of brood (sealed and unsealed) for three days kills *T. clareae* adults (Woyke, 1985). Dusting of wheat flour on bees also hinders the mite attachment to bees and contributes towards its control. Natural resistance mechanism in terms of 'lycurgen behaviour' by *A. dorsata* (Burgett *et al.*, 1990) and 'removal behaviour' by *A. mellifera* also checks low level of infestation.

6.3.2 *Endoparasitic mites*

Such mites are internal parasites and generally found in the respiratory tract of insects. These mites can be observed only after examining the internal body organs of bees. About fifteen species are reported to parasitize Orthoptera, Hymenoptera, Lepidoptera and Hemiptera (Sammataro, 1997). These parasitic mites can be observed only after examining the internal body organs of bees. *Acarapis woodi, A. externus, A. dorsalis* falls under this category.

6.3.2.1 *Acarapis woodi* (Rennie)

This is commonly called as the honeybee tracheal mite (HBTM) which live inside the tracheae and air sacs of adult honeybees of *A. cerana, A. mellifera* and *A. florea*. *A. woodi* infestation occurrs with in two days of emergence of worker bees and 52-60 per cent bees are found infested with in a week of initial infestation (Atwal, 1967; Dhaliwal *et al.*, 1974). About ninety per cent losses have been reported in USA (Sammataro *et al.*, 2000).

A. woodi is a small parasitic mite that infests the airways of the honeybee. The first known infestation of the mites occurred in the British Isles in the early 20th century. It was first observed on the Isle of Wight in 1904, the mystery illness known, as Isle of Wight Disease was not identified as caused by a parasite until 1921. It quickly spread to the rest of Great Britain. It was regarded as having wiped out the entire bee population of the country and dealt a devastating blow to British beekeeping.

Mature female acarine mites leave the bee's airway and climb out on a hair of the bee where they wait until they can transfer to a young bee. Once on the new bee, they will move into the airways and begin laying eggs.

Biology

The entire life cycle is spent in the respiratory tract of honeybee, except for brief migratory periods. There are only three stages in the life cycle: egg, larva and adult. However, the mite has nymphal instar that remains inside the larval skin (Lindquist, 1986). Each female lays 5-7 eggs, which requires 3-4 days to hatch. Development of males and females is completed in 10-11 and 14-15 days, respectively. All stages remain with in the tracheae while adult female searches for a young bee or a drone and enters the tracheae through spiracles.

Symptoms

The tracheae of severely infested bees have brown blotches with brown scabs or crust like lesions and may appear black. A tracheae mite infestation shortens adult bee life, affects flight efficiency as a result of atrophied flight muscles and reduced the bee ability to thermoregulate. Large number of crawling bees in front of infested colony, 'K' winged condition due to separation of fore and hind wings (Goyal and Kandoria, 1988), falling of bees during play flight (Desh Raj *et al.*, 1991), yellow droppings in front of hives and hive parts and presence of sluggish bees with distended and shining abdomen are some of the other symptoms associated with *A. woodi* infestation.

Disease caused by tracheal mites is called Acarine disease or Acariosis, named after its scientific name *Acarapis woodi*. Tracheal mites can't be seen with the naked eye. Confirmation of infestation requires microscopic examination. Infected colonies are most likely to show signs of infestation in spring and fall when mite populations reach their peak.

In cases of severe infestation, large numbers of adult bees crawl in front of the hive incapable of flight and coordination. Many adults extend their wings outward, often described as "K-Wing". Infected adult bees may be lethargic. Even on good days, a high proportion of bees are in the hive instead of flying. Colonies are slow in population build up in the spring and lack vigour.

Control

Use of menthol crystals (50g) for 2-4 weeks (Kshirsagar *et al.*, 1980; Southwick, 1988; Sammataro *et al.*, 2000), formic acid (Sharma *et al.*, 1983; Hoppe *et al.*, 1989) as fumigants in infested colonies provide effective control against *A. woodi*. Five to seven fumigations with folbex strips impregnated with potassium nitrate and 500mg chlorobenzilate at weekly interval is also recommended to control *A. woodi* infestation (Atwal, 1971; Shah and Shah, 1988).

Several resistant lines of bees have also been developed, commercially reared and sold to beekeepers (Danka *et al.*, 1995; Lin *et al.*, 1996).

Acarine mites are commonly controlled with grease patties (typically made from 1 part vegetable shortening mixed with 3-4 parts powdered sugar) placed on the top bars of the hive. The bees come to eat the sugar and pick up traces of shortening which disrupts the mite's ability to identify a young bee. Some of the mites waiting to transfer to a new host will remain on the original host. Others will transfer to a random bee–a proportion of which will die of other causes before the mite can reproduce.

Before applying chemical controls, one should closely examine the colonies for signs of HBTM infection. Adult bees crawling in front of the hive, lack of vigour, displays of K-Wing are some of the symptoms associated with *A. woodi* infestation.

Increased incidence and severity of infestation is worsened when bees are stressed. Stress factors may include presence of other diseases, poor forage availability, too many colonies placed in one location causing interference and robbing, lack of feed and pollen, poor beehive equipment, inclement weather, etc. Other preventative measures include origins of swarms should not be collected; reduce movement of colonies and avoid high density areas; don't introduce colonies from infected areas or that have not been tested into the beekeeping operation.

6.3.3 Phoretic Mites

These mites use bees as carriers to move from one place to another. When in large numbers, they adversely affect the foraging behaviour of bees *e.g. Neocypholaelaps apicola, N. indica, Afrocypholaelaps africana* (Seeman and Walter, 1995) and *Aceria litchi* (Waite and McAlpine, 1992; Waite, 1999). They feed on pollens and leaves of tropical and subtropical trees.

6.3.4 Scavengers

These mites are found in the bee nest or inside the comb cells. They are found in the stored pollen and honey and other decaying organic matter. These mites are also associated with stored products and dust. The examples are *Austroglycyphagus geniculatus, Tyrophagus longior, Acarus siro, Rhizoglyphus* sp. In addition to this, Hypopus of several mite species *viz. Kuzinia* sp., *Sennertia* sp., *Vidia* sp. adhere to active bees and cause nuisance when their population is high.

6.3.5 Predatory Mites

Predatory mites are also reported from beehive, which feed on parasitic, phoretic and scavenger mites and provide some control of these harmful mites. The families of these predatory mites are Cheyletidae, Cunaxidae, Bdellidae and Phytoseiidae. Among these, *Acaropsis sollers, Cheletophyes ruttneri, C. shendei, C. deodikari, C. orientalis, Cheyletus malaccensis* are important mites.

Chapter 7

Water Mites

The 'true' water mites, the Hydracarina, include about 5000 mainly freshwater species in more than 40 families. Water mites are members of the Cohort Parasitengona from the Suborder Prostigmata (Order Acariformes). Parasitengones are characterized by a complex life cycle. These water mites (Acari: Hydracarina) are predaceous as adults and deutonymphs and parasites as larva. The larva is mainly parasitic on insects with a wide range of specificity (Baker, 1996; Smith, 1990; Goldschmidt *et al.*, 1999). Larvae of water mites parasitise all orders of aquatic insects with the exception of the Megaloptera, Neuroptera, and Ephemeroptera. Post-larval stages feed on various stages of aquatic small arthropods, including copepods and cladocerans, insect eggs, and sometimes rotifers and nematodes (Smith and Cook, 1991; Proctor 1997).

Water mites inhabit all types of aquatic habitats, from hot springs to treeholes. Water Mites live in marshes, ponds, and lakes. They like shallow, still water with lots of plants. One family, the Pontarachnidae, has invaded nearshore marine waters. Some taxa are excellent swimmers while others crawl on or in the substrate. Many genera, particularly early derivative ones, are characterised by bright orange or red colouration. These mites also tend to be distasteful to fish (Kerfoot 1982).

Chemoreception and mechanoreception are used in prey selection and in the mating behaviour of several water mite species (Proctor and Pritchard, 1990). Unionicola use chemoreception and light in the selection of mussels as hosts (Dimock, 1988). Few parasitic mites on aquatic insects are reported from India. These are considered as nuisance for those aquatic insects, which constitute an important fish food spectrum.

7.1 Distribution

Although earlier researchers considered the Hydracarina to be grossly polyphyletic, studies of larval morphology, genitalia, sperm ultrastructure, and the presence of unique trigger-hair/gland structures (glandularia) support the monophyly of nine superfamilies: Hydrachnoidea, Eylaoidea, Hydryphantoidea, Lebertioidea, Hygrobatoidea, Arrenuroidea, Sperchontoidea, Hydrovolzioidea, and Stygothrombidioidea (Harvey 1998; Walter and Proctor, 1999).

The red mites are perhaps the best-known water mites since they are easily visible. They reach sizes up to 6 mm. There are several similar genus. All water mites are predators. They feed on all kinds of Crustaceans and larvae of water insects. The pair of pedipalps grabs the prey and the fluid content of the victim is being sucked out via piercing mouthparts.

Water mites' ancestors were land dwelling. They don't possess any gills. But they are well adapted to a life under water. They have trachaea, a system of very fine tubes that transport the oxygen throughout the body (like insects have). The ends of the tubes are covered with a thin membrane so oxygen can be extracted from the water by diffusion.

Water mites are familiar to everyone who ever observed pond water. The larger species are clearly visible as fast swimming sometimes-deep red spider like creatures. Determination of, to what genus a water mite belongs, is not very hard but species identification is much more difficult. Species can be identified on the basis of the reproductive organs, the eyes or palps.

Davids and7 Schoots (1975) and Lanciani (1979, 1982) demonstrated the influence of water mite parasitism on non-pest insect population while Lanciani (1986) and Smith and McIve (1984) showed similar adverse effects on mosquito numbers.

7.2 Lifecycle

The life cycle of water mites is quite complicated but interesting too and there is still a lot to be discovered. The development differs in each family and often in each species. The eggs develop into larvae that often live a parasitic life. Some attach themselves to larvae of dragonflies or mosquitoes, others attack adult water insects like water beetles. In some cases these parasitic stages are absent.

After mating in late spring or early summer, female mites lay eggs on stones or plants in the water. Eggs are red and one female lays up to 400 at one time. Eggs hatch in one to six weeks, depending on the species and temperature.

Mite larvae hatch with only six legs. Immediately after hatching, they swim around looking for an aquatic insect. Water mite larvae are parasites and they need a host to live on. Some insects chosen to be hosts include dragonfly naiads, damselfly naiads, fly larvae (including mosquitoes), true bugs and stone fly nymphs. One insect may carry around up to 20 mite larvae.

Water mite larvae munch on their host while it carries them. When a mite larva has grown enough, its exoskeleton (outer shell) becomes bag-like. The larvae stays inside the bag and changes to a nymph. As it changes, the "bag" stays attached to the host.

When the nymph is ready, it leaves the host. The nymph has eight legs and looks a lot more like an adult water mite. Water mite nymphs eat the same foods as adults do, including small insects, aquatic worms, and other mites (even their own kind). Nymphs continue to eat and grow, until they are ready to go into their fourth stage (egg, larvae, and nymph are the first three stages).

Each nymph finds some algae, an aquatic plant, or another object to attach to. The nymph shrinks inside another "bag" for about a week while it changes to an adult. Predators of adult water mites include aquatic insects, crustaceans (such as water fleas and scuds), hydras, and small fish.

In the larval stage, when attached to its host, the mite may have a problem if the host leaves the water. If this happens, the mite can pop out of its bag, or hold on, hoping its host returns to the water. If the host doesn't return, or the larva drops off on land, it dies.

7.3 Families and Commonly Occurring Genera

Work on taxonomy, ecology and behaviour are documented by Cook (1976; 1980; 1986), Harvey (1998), Goldschmidt (2001; 2002), Springer and Gerecke (1992) and Valdecasas (2001). Families associated with water mites are Galuminadae, Hydrachnidae, Midiopsidae, Limnocharidae (Rhyncholimnochares), Hydryphantidae (*Protzia)*, Hydrodromidae (*Hydrodroma*), Rhynchohydracaridae (*Clathrosperchon, Rhynchohydracarus,* Santiagocarinae), Sperchontidae (*Sperchon)*, Lebertiidae (*Lebertia*), Oxidae(*Frontipoda*), Torrenticolidae(*Monatractides, Neoatractides, Torrenticola, Pseudotorrenticola*), Limnesiidae (*Limnesia, Neotorrenticola, Neotyrrellia, Protolimnesia*), Omartacaridae (*Omartacarus*), Hygrobatidae (*Atractidella, Atractides, Corticacarus, Hygrobates, Paraschizobates*), Unionicolidae (*Koenikea, Neumania, Unionicola*), Aturidae (*Albia, Aturus, Axonopsalbia, Axonopsella, Axonopsis, Diamphidaxona, Frontipodopsis*), Mideopsidae (*Mideopsis)*, Krendowskiidae (*Geayia)*, Arrenuridae (*Arrenurus, Stygarrenurus*).

The Halacaridae (sometimes split into several families) is a group of entirely aquatic mites from the Suborder Prostigmata of the Order Acariformes. The majority of the species are marine, but there are many brackish and freshwater taxa as well (Bartsch 1996). Some are predatory, some algivorous and some appear to be parasitic (*e.g.* Astacopsiphagus in the gills of the crayfish Euastacus spinifer).

The Pezidae, another family of halacaroid mites, can be differentiated from the Halacaridae by their possession of numerous small dorsal platelets bearing 10-14 pairs of stalked setae; 'typical' halacarids have at most 6 pairs of small dorsal setae (Harvey 1990).

The Hydryphantidae are commonly known as red water mites, which have a soft, papillate or lined integument. Coxal plates are divided into four groups and lateral eyes are located in distinct capsules. The larvae of several genera parasitize mosquitoes, black flies, biting midges and horse flies whereas nymphs and adults are free-living predators feeding on mosquito eggs (Mullen, 1975; Smith, 1983; 1988; Smith and Oliver, 1986). *Thyas barbigera* Viets completes its life cycle in two years and on an average 45 eggs are laid by each female (Mullen, 1977).

7.3.1 *Arrenurus*

Arrenurus mites grow up to three millimeters long and are usually brown or greenish. Males are bigger than females. *Arrenurus* mites stay near the bottom. They can swim or crawl to get where they are going, but when they relax; they sink back to the bottom. *Arrenurus* mites are only active when there's daylight. The males of the mites of the genus *Arrenurus* have modified rear ends. Every species has its distinct shape. Their tough carapace has a texture of fine punctures. The females of *Arrenurus* lack the tail. These mites show colours from grey to brown and blue.

7.3.2 *Unionicola*

The mites of this genus have very long legs. The shape and size of the legs differ according to the life style of the mite. Swimming species have fine hairs on their legs while in bottom dwelling species the legs are modified to crawl.

7.3.3 *Torrenticola*

It belongs to the family Torrenticolidae.

7.3.4 *Hydrarachna sp. (Hydrachnidae)*

Mites from the genus *Hydrachna* are very common. When examined under the microscope the skin appears armoured with very fine spikes. It is recorded from *Stenopsyche haimavatika* pupae on the intersegmental areas of abdomen. Insect larvae are not found to be parasitized by this mite. The per cent parasitism is found to be 11.1 to 73.3. The average number of mites per host varied from 1.6 to 10.3. Incidence of parasitic mite is synchronized with the insect population, higher the host population higher the mite incidence. Under heavy infestation, pupal death is reported.

7.3.5 *Histiostoma sp. (Histiostomatidae)*

It is found to attach on the undersurface of elytral bases joints of sternal plates and legs of the semi aquatic beetles, *Paedurus nigricornis* and *P. basalis*. The per cent parasitism of this mite is recorded upto 60 and 68.8 per cent in *P. nigricornis* and P basalis, respectively. The infested beetles are less active than healthy ones.

PART II:
Acarological Techniques

Chapter 8

Collection and Extraction of Acarines

Separating mites from samples is an essential part of any survey done either for population dynamics and biology of particular species or its management. Extraction procedure should be economically viable and without causing significant loss or damage to mites to be collected. Sampling method besides yielding reliable estimates of the mean population density and its standard error, should balance the precision of the acquired information with the cost of time. Sampling strategies are usually based on the direct count of mites. Although it is extremely time consuming, it cannot be avoided as a first step towards the development of shortcut sampling methods based on indirect measurements (Hart and Fain, 1987; Braun *et al.*, 1989; Yaninek *et al.*, 1991). According to Nachman *et al.* (1993), surveys based on well designed sampling plans are more general, therefore, less sensitive to violations of underlying assumptions than the indirect methods that normally rely on specific calibration functions transforming the indirect measurements to estimates of mean population density.

Successful extraction depends on the sound knowledge of the mite behaviour and acquaintance with its habits, habitats and symptoms (Henderson and McBurnie, 1954; Henderson, 1960). Majority of the acarines are photonegative in nature. Tetranychid and Tenuipalpid mites are mainly found on the under surface of the leaves of agricultural crops. Oribatid mites are concentrated in litter whereas nematophagous mites can be extracted from the root zone area of crops in soil. Likewise, storage mites are best collected from the left over material in grinding mills and stored go downs.

Necessary materials for collection of mites include polythene bags, paper envelops, rubber bands, labels, brush, vials, preservatives and pocket/hand lens.

8.1 Housedust Mites

8.1.1 Dust Sampling

Dust sampling is one of the simplest methods for the isolation of mites from furnishings but this provides only relative figures not the total figures. Van Bronswijk (1973, 1981) and Fain *et al.* (1988) described different methods of mites' isolation from dust samples either through floatation or vacuum cleaning. Repeated vacuum cleaning of carpet or upholstery with in a short time span, gives good collection of housedust mites (Bischoff and Van Bronswijk, 1986). Both live and dead mites are collected by this method. Colloff (1987, 1988) suggested that two or more comparable samples could be taken at the same time from carpets or mattresses (large textile objects), which are uniformly infested by mites.

8.1.2 Heat Escape Method

This method is based on the mites' sensitivity to desiccation. Dry air is generated by careful heating of one side of an object that favours the mites to migrate to the other side in order to escape the influence of heat. A metal hotplate with a temperature indicator and controllable electrical resistance is used. An initial temperature of 26-28°C to an end temperature of 65°C with test duration of thirty minutes resulted in 65-78 per cent mite collection (Bischoff, 1988). This method has a limitation that it can not be applied to heat sensitive furnishings.

8.1.3 Mobility Test Method

Representative parts of a textile object is covered by adhesive films, weighed down with a glass plate and protected against light exposure. The covering plate ensures good contact between the support and the adhesive film to allow the living mites to move easily from the fibres of the textile object (Bischoff *et al.*, 1992). After 24h, adhesive films are carefully removed and mites are counted under microscope.

8.1.4 Collection Method for Mounting

Colloff (1989) described a rapid method for collecting mites from dust samples for mounting purposes (Figure 8.1). From suspended dust samples in lactic acid, mites are extracted via a fine pipette into a watch glass containing a few drops of distilled water. Excess dust particles and lactic acid are removed from watch glass via pipette with a tip smaller than the smallest mites. Mites are frozen in water in watch glass at–20°C. The ice containing mites are then transferred from watch glass to siliconeized microslide (ordinary glass slide coated with silicone solution and dried properly), melted and evaporated on a hot plate at 55°C. Water based mounting media and cover slip can then added in a normal way.

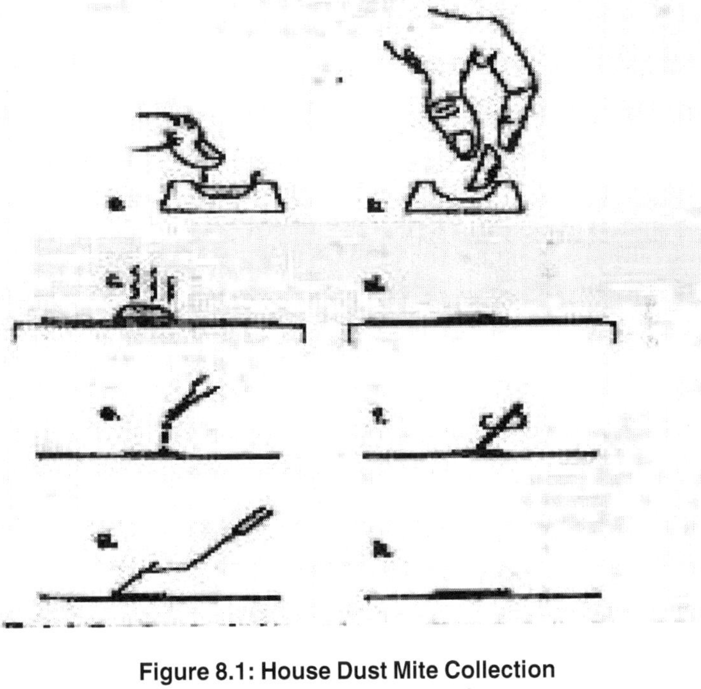

Figure 8.1: House Dust Mite Collection (Colloff's method,1989)

8.1.5 Mite Detection in Water

Vacuum mattresses, carpets, sofas, and chairs and examine the dust collected for mites. A simple method for detecting the presence of mite is to put a small amount of dust on the surface of water and examine it under 20X magnification. Live and dead mites will remain on the surface of the water.

8.1.6 Ultrasonic Technique

Furumizo (1975) reported a sieving and flotation technique. A sample of dust (0.05 g) is placed in 30 ml of saturated NaCl solution and 5 drops of detergent are added. The dust is teased apart and subjected to ultrasonic treatment for 20 minutes. The suspension is rinsed through a 45 micron mesh sieve, stained with crystal violet and examined under a stereo microscope.

8.1.7 Mite Detection through DIN-A4 Paper

Dust mites are attracted to white DIN-A4 sheets of typewriter paper. These are placed where the dust mites are suspected. Paper is examined in the mornings and evenings by holding the paper over a lamp. The mites cast a shadow.

8.2. Parasitic Mites

Different techniques applied to collect parasitic mites attached on domestic animals, wild animals, insects, birds are given below under separate headings. Unattached parasites in nests and other nidicolous species may be collected by simple heat desiccation techniques.

From Birds

8.2.1 Direct Examination

The birds with rough leg skins, sparse feathers and lazy birds should be picked up for probable mite infestation either through naked eye or hand-lens. If mite infestation is noticed in legs, washing with soap solution in an enamel tray will dislodge the mites from legs, which are collected in tray and can be transferred to tight fitting vials for detailed examination under microscope. Few feathers (2-3) from each bird showing mite infestation symptoms can be plucked and collected in polythene bags for later examination in laboratory.

8.2.2 Fumigation Method

For living birds, "Fair Isle" apparatus is used. Bird is enclosed in a cylinder with its head and neck protruding through a hole in the upper lid. Chloroform vapour is pumped into the cylinder and the bird is encouraged to flutter so that anaesthetized parasites drop off on a suitable coloured tile, which forms the base of apparatus. An oiled silk sleeve around the neck prevents the chloroform reaching the birds nostrils. Mites can also be separated from birds by a fumigation technique (McGarry and Trees, 1991) described under subheading 8.2.7.

8.2.3 Flushing Technique

This technique was originally proposed by Faccini *et al.* (1987) for *Raillietia* species (ear mite) in cattle, which was later, used for *Psoroptes* species, ear mite of sheep and goats (Leite *et al.*, 1989; Faccini and Costa, 1992). Flushing the ear canals by 50 ml of water provides good collection of ear mites than by swabbing or otoscopic examination. The side effect of this technique is a transient loss of equilibrium in some animals.

8.2.4 Nasal Examination

The dead birds can be examined by this method for mite extraction. Alcohol (70 per cent) is injected into one nostril with the help of a syringe and solution coming out from second nostril is collected in a beaker or wide mouthed vial, which can be then examined for presence of mites.

8.2.5 Berlese Funnel Method

Bird feathers are kept in a sieve, which is then placed on a funnel. The distal end of a funnel is dipped into a vial containing water or 70 per cent alcohol (Figure 8.10). With the help from a light source in the form of an electric bulb, kept above the sieve, mites are collected in vial after 24h. A thin layer of petroleum jelly is applied on the vial tip to prevent mites from escaping.

From Domestic Animals

8.2.6 Brushing and Combing Method

Brushing and combing of host animal body over a shallow dish is one of the easiest methods to collect parasitic mites. Washing by detergent solution is also helpful in dislodging the mites.

8.2.7 Fumigation Method

Pet animal is placed in a plastic bag containing a small cotton boll soaked in chloroform for approximately ten minutes with only its head protruding. Mites are then washed out of the plastic bag with 70 per cent alcohol into collecting bottles (Mc Garry and Trees, 1991).

8.2.8 Skin Scrappings

Another productive method is to scrape the skin of the animal with scalpel at various locations of the body separately *viz.* inside and outside the pinnae, dorsal surface of the head and trunk, both pairs of limbs including their bases, ventral surface of the trunk with special emphasis on the genital region and the tail. Recording of the sex, age and symptoms on host is must.

During survey the scrappings from animal are collected in large petri dishes (dia. 9 cm), which are then transferred to properly labeled polythene bags and tightly tied. In laboratory, these scrappings are transferred to petri dishes and petroleum jelly is applied on the margins of petri dishes to prevent the escape of the mites. These scrapings are examined either directly under stereo-binocular microscope for presence of mites or after boiling them in 5 per cent potassium hydroxide solution for few minutes and centrifuging lightly (Sediment will contain mites). Small amount of scrapping is placed on micro slide with two-three drops of water/glycerine/potassium hydroxide solution under the cover slip for examination.

8.2.9 Small Wild Mammals

Small wild mammals (rats, squirrels *etc.*) are trapped in wire cages. Bats can be collected by mechanical attacks. These animals are killed with chloroform in laboratory in separate jars containing white paper at the bottom. Any mite detached from the dead animal's body can be collected on white paper. Then the animal is examined with naked eye and hand lens for mite infestation. Different parts of the body are brushed separately and in another jar/beaker mites are collected on white paper. Subsequently, washing the dead animal thoroughly with detergent to dislodge the undetected mites from host body. This water sample is then transferred to counting dish/small petri dish for counting of mites under microscope.

From Insects

8.2.10 Lycopodium Dusting

Insects are dusted with safranin dyed *Lycopodium* powder before releasing them in tray for 24 hours. Tray is covered properly to prevent the escape of insects. Examination of residual powder in tray will show the presence of mites, which can be collected, with the help of moist pick. Insects can be directly inspected under microscope for any attached mite remaining on insect body. Cheyletid mites are usually found on the ventral side of insects, at joints and in beetles under elytra. Mites can also be collected from insect wings.

8.2.11 Adult Insect Examination

Insects including honeybees are collected and kept in BOD for two hours which sub due their activity. They are then examined with the help of hand lens and later released in the field/hive.

8.2.12 Dead Insect Examination

Hand shaking of dead insects in beaker or any wide mouthed bottle/vial containing alcohol will dislodge the mites from insect body. In place of alcohol, one can use ether, detergent solution or hot water.

8.2.13 Bottom Board Insertion in Beehives

Examination of bottom board inserts is a useful method for detecting the presence of mites, measuring acaricide efficiency, estimating population growth and size of mites. Bee mites are collected on white plywood boards thinly covered with vegetable oil, on plastic sheets smeared with vaseline (Accorti *et al.*, 1986) or on simple white paper placed at the bottom.

8.2.14 Hive Debris Examination

Regular examination of hive debris on bottom board (Calatayud and Verdu, 1993) or on white thick sheet will reveal the dead mites, which confirms the presence of mites in beehive (Figure 8.2). It is advisable to change the sheet everyday.

8.2.15 Hive Brood Examination

Worker and drone brood can be removed from cell after uncaping and examine it directly under microscope. Brood cells are

Figure 8.2: Hive Debris Examination

uncapped; brood is taken out and examined directly for presence of mites in brood/with in the cell with the help of hand lens.

8.2.16 Sampling with Acaricides

An acaricide strip is inserted into the colony for one to two days. Dead mites can be trapped on paper either smeared with petroleum jelly or kept on bottom board under 2-4 mm mesh screen. Two drops of fumigants on to a strip of paper previously impregnated with potassium nitrate is burned with in the hive. Mozes-Koch *et al.* (2000) found this method very effective as dropped mites are collected and counted one hour later.

8.2.17 Floatation Method

Placing the debris collected from 8.2.10 and 8.2.13 method, in a jar containing 98 per cent alcohol will show floating mites on the surfaces from where they can be removed and examined.

Randomly selected adult bees can be collected in the apiary and placed in alcohol water solution (50 per cent). In case ethanol or rubbing alcohol (propanol) is not available, bees can be temporarily placed in windshield wiper fluid containing methanol. Tracheal mite infestations in an apiary are often un-equally distributed.

Typically, there are "hot spot" infestations. For this reason, each colony in the apiary must contribute equally to the bee sample being collected.

A sample of bees from an apiary, regardless of the number of colonies, must contain at least 75 bees in order to detect a general tracheal mite infestation of 2.5 per cent or higher in the apiary.

A. woodi is an obligate, endo-parasitic mite affecting the trachea or breathing tubes of the thorax of adult honeybees. Using a dissecting microscope in the laboratory, adult bees are placed on their back, decapitated and the frontal section (collar) of the thorax is removed with a surgical blade or pulled with fine forceps. The major tracheae are exposed and can be examined. Healthy, non-infected tracheae appear milky-white in color and almost transparent. Infected tracheae are mottled, sometimes bronzed in appearance, and with darkened patches. Higher magnification sometimes allows for the identification of adult mites. Light infestation generally involves one tracheal tube affected by mites (mono-lateral) while heavy infestation involves the damaged tissue of both tracheal tubes (bi-lateral).

8.3 Phytophagous Mites

These are normally found on the ventral surface of leaves but when population is very high, mites move to the upper surface of the leaves as well. A good number of mites are found on leaf lamina, in angles formed by major veins, flowers, buds, axils of leaves, twigs, stems or under barks.

Brevipalpid and eriophyd rust mites can be collected by examination of half-inch square grid areas on leaves by means of a linen counter. Eriophyd mites are often found on malformed or deformed plant parts. Mite galls are of open type contrary to insect galls, which are of closed types.

Plant parts showing symptoms of speckling, browning and other discoloured patches on leaves, galls, erineum patches, crowded buds *etc*. have to be collected and put in individual polythene bags or paper envelops with proper labels. If storage of samples is necessary beyond 2-3 days, paper envelops should be preferred for leaf collection. Polythene bags containing leaves show mold growth due to high moisture. In absence of paper envelops, plant materials in polythene bags can be refrigerated. Identification of host plant is also important hence herbarium should be maintained in a laboratory.

Cochran (1977) and Sabelis (1985) have reviewed the various sampling techniques for plant mites. Numbers of specialized collection techniques (Zacharda *et al.*, 1988) are followed for the collection of plant mites from foliage, which are summarized briefly:

8.3.1 Hand Picking

This method is one of the oldest, strenuous methods yet frequently followed as it provides rich collection of plant mites. Infested leaves as such can be plucked from plants and collected in polythene or paper bags with proper labels. Tetranychid mites can be best collected by randomly plucking leaves from plants or trees. In a 3×2 cm area, samples should be collected at least from ten plants. Collection of male specimen is essential for identification of tetranychid species; therefore, necessary efforts should be made to get males in the collection. A fine sable-hair brush or a pick made from bird's feather (Figure 8.3) moistened with 70 per cent alcohol can be used for mite picking before releasing them in vial containing alcohol (70 per cent). Tydeid mites being smaller in size can be best collected by hand picking.

8.3.2 Sweeping

Sweeping with a fine net on low herbaceous plants, grassland yield large number of mites entangled on the inner surface of the net wall (Figure 8.4). These mites can be picked up with a pick or sable hairbrush.

8.3.3 Beating

In this method, a white enamel tray with cotton pad is placed under the plant canopy. Plant parts are beaten with a wooden rod, which results in dislodging of mites, and their subsequent fall on the cotton pad. These mites are picked up by fine brush and placed into preservative. Very small sized mites can be lost in this procedure.

Figure 8.3: Bird Feather Pick

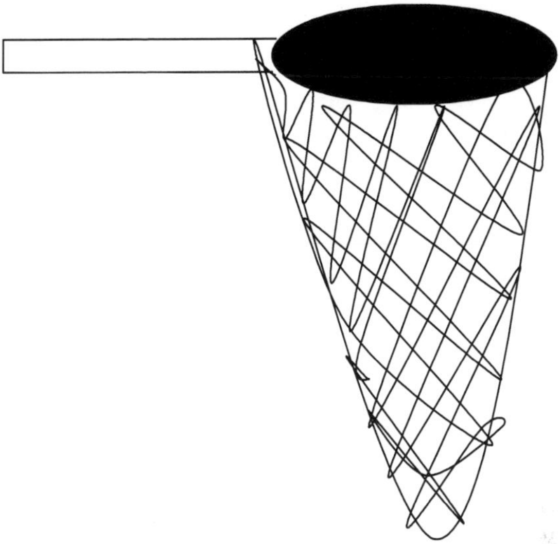

Figure 8.4: Sweeping with Net

Instead of cotton pad, half-inch mesh screen covering a large tray gives better results (Krantz, 1978).

8.3.4 Aspirator

Singer's (1964) aspirator (Figure 8.5) is quite convenient for the collection of plant mites as it collects the mites from its habitat directly into the preservative (70 per cent alcohol), eliminating the necessity of handling the specimen. This method is not useful when live mites are to be collected.

8.3.5 Stratified Sampling

Sampling with three vertical strata, each stratum being one third of the foliage height (Figure 8.6), yields the most reliable estimates of within plant population density, because it accounts for the tendency of the mites to aggregate in the upper parts of the foliage and at the same time minimizes the sampling effort for a desired level of precision. The two leaves samples from each stratum of a plant or from sub-sample of plants provide estimates of mite density both at per leaf and per plant basis (Nachman *et al.*, 1990). If lobed leaves are there, collection of leaflet as a sampling unit will reduce

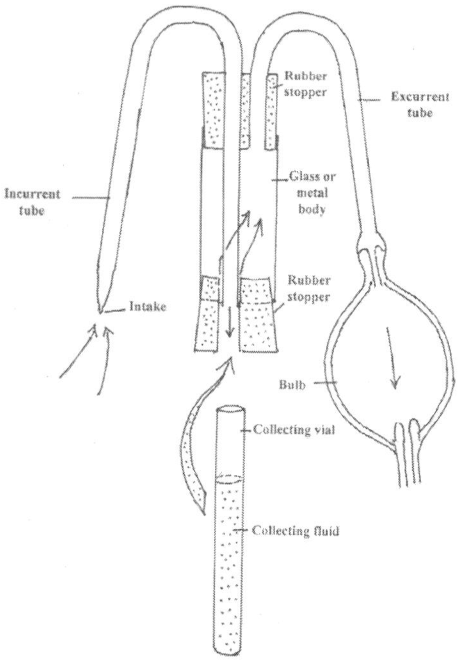

Figure 8.5: Aspirator (Singer, 1964)

the sampling time as variations with in leaf are smaller than the variations between leaves.

8.3.6 Sampling from Large Area

After stratifying the field horizontally into sub plots of known number of plants, one plant is selected at random from each sub plot on each sampling day and stratified sampling is done (Nachman *et al.*, 1993; Skovgard *et al.*, 1993). To reduce the error, one should either increase the number of primary units (plants) or the number of secondary units (leaf) per primary unit and/or the number of plants used for estimation of stratum sizes.

Information on the plant size and structure with respect to plant height, number of main stems, number of leaves per stratum, leaf height from ground and length of leaf is also considered as essential. Generally, positive correlation between mite density and plant height is observed which implies that high mite densities are associated with large plants in most cases.

Figure 8.6: Stratified Sampling

8.3.7 Vacuum Sampling

Mites can be collected with a petrol driven sampler and then can be transported to laboratory in small-ventilated vials with fresh leaves as food and moist sponge to maintain humidity (Wallace, 1972).

8.3.8 Photographic sampling method

Asquith (1965) pioneered the use of photography for sampling and counting mites and later, Sircom (2000) described four advantages of *in situ* photography:

1. Large number of samples can be collected at one time and counting can be done later.
2. The record is permanent so recounting/re-examining is possible.

3. Information can be obtained on the small-scale spatial distribution of the mites
4. Neither plant material nor mites are removed from the population thus there is minimal disturbance in the population dynamics.

Sircom (2000) used a camera with macro-lens, orange filter and data back system (Figure 8.7) for photographic sampling of plant mites. For rapid and accurate photography, an aluminium frame was attached to the camera with a plexiglass bracket to hold the leaf. Photographic counts of *T. urticae* and *P. persimilis* were 92 and 98 per cent, respectively, of direct counts.

8.3.9 Leaf-washing Apparatus

Jedlickova (1997) gave the leaf-washing apparatus (Figure 8.8) for the recovery of mites. The apparatus consisted of the following.

(1) A plastic hemispherical or commercial rinse dish (dia. 28cm) used for fruit washing, with holes (4 x 4 mm) in the bottom. The dish serves as a reservoir for placing the whole sample.

(2) A rough sieve (mesh 2mm) for catching dirt and bigger leaf fragments and/or bigger arthropods if present. Standard laboratory sieves with an appropriate mesh size may be used.

(3) A middle washing sieve (mesh size 0.3-0.5 mm) for catching smaller leaf fragments, bigger trichomes, dust and/or arthropods.

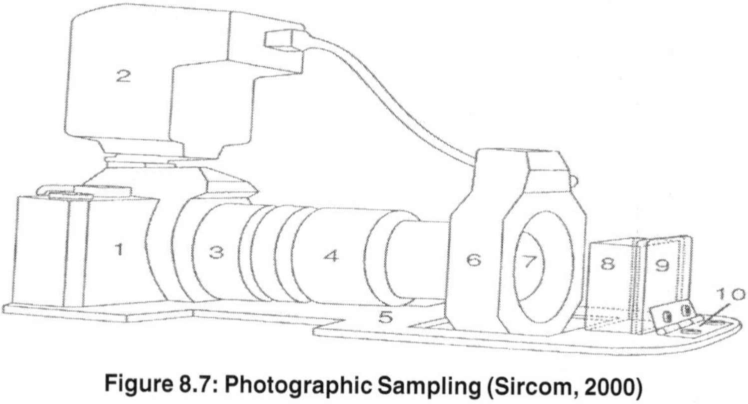

Figure 8.7: Photographic Sampling (Sircom, 2000)

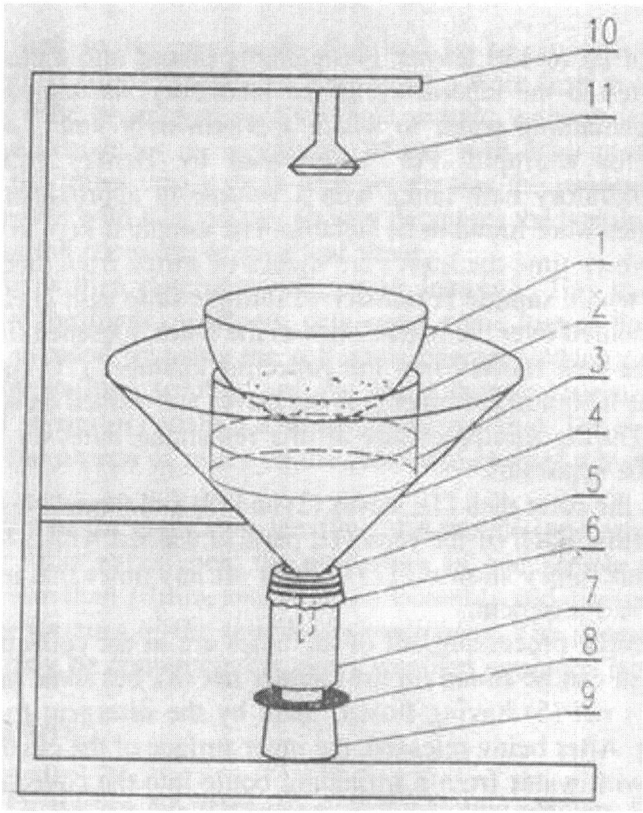

Figure 8.8: Leaf Washing Apparatus (Jedlickova, 1997)

(1) Plastic hemispherical rinse dish, (2) Rough sieve, (3) Middle washing sieve, (4) Funnel, (5) Circular nylon net, (6) Rubber rings, (7) Plastic collecting chamber, (8) Circular net, (9) Plastic pot with cut off bottom, (10) Portable laboratory stand and (11) Laboratory hand-controlled shower.

(4) A funnel with a diameter of approximately 60cm.

(5) A circular nylon net (mesh size approximately 0.16mm) with a diameter approximately 4cm larger than the diameter of the plastic chamber (7). In the middle of the net there is a circular opening with the same diameter as the lower funnel tube. The net prevents the possible escape of mites that could float on the detergent foam during the waste washing.

(6) Rubber rings for fastening the nylon net (5) to the plastic chamber (7).

(7) A plastic collecting chamber made from a wide-necked plastic bottle (200-250ml) at least 11 cm high, by cutting off the bottom (in the apparatus the top); the neck should be at least 4cm in diameter and provided with a thread.

(8) A circular net of the same material as the net (5), with a diameter larger than the neck of the chamber (7). This net serves to catch mites and transfer them to a microscope or to a vial where the mites can be deposited for later study. The preparation of a large number of such nets is recommended.

(9) A plastic pot with the bottom cut off, approximately 5 cm high, with a bottom diameter the same as the outside diameter of the neck (serving as the bottom) of the collecting chamber. If the plastic pot has a thin wall, strengthen it with adhesive tape. Instead of a plastic pot, the twist cap (stopper) from the bottle (7) may be used; its centre is carved out. The pot serves to fasten the net (8) to the collecting chamber (7) and as a base for the apparatus if a portable laboratory stand is not used.

(10) A portable laboratory stand for fastening the apparatus; if the pot (9) is used then portable laboratory stand is not required.

(11) A laboratory hand-controlled shower.

Construction *(Figure 8.8)*

The nylon net (5) is placed on top of the collection chamber (7) and fixed by the rubber rings (6). The net (8) is fixed on the lower part of the collecting chamber (7) using the plastic pot (9) with the cut off bottom upwards. The funnel outflow tube is inserted into the circular opening of the nylon net (5). Then both washing sieves (2 and 3) are put into the funnel (4) (the rough sieve over the middle one). The rinse dish (1) is placed over the rough washing sieve (2). The apparatus is then placed into the laboratory sink in the plastic pot (9) in such a way that water outflow is guaranteed. The equipment can be stabilized by the laboratory stand (10).

8.4 Plant Predatory Mites

8.4.1 *Anystidae*

Sweep net provided with a collecting bottle at the bottom was used by Lange *et al.* (1974) to collect adult mites. They also trapped adults from trees with belts open only at the entrance. *Anystis baccarum* egg clusters are collected from the pine and oak litter through sieving. Anystid mites are easily collected from foliage because of its bright red colour, characteristic figure of eight and whirling running pattern (Muma, 1975).

8.4.2 *Bdellidae*

As Bdellid mites are usually encountered under bark or on fallen branches, Currie (1934) used wooden boards as mite traps in the field and with the help from aspirator, 1000 mites per person per hour were collected from these boards. Wallace (1972) developed a portable, motor operated sucking machine to collect these mites.

8.4.3 *Phytoseiidae*

Stratified sampling and photographic sampling method described under 8.3.5 and 8.3.8, respectively, provides reliable estimates of phytoseiid population. Predatory population is counted per leaf as compared to phtyophagous mite (except eriophyid) population as per sq.cm.

8.5 Soil Mites

Soil mites can be collected from litter and soil around the tree/ crop. A number of methods have been developed for the extraction of invertebrate animals from soil and litter samples (Macfadyen, 1955; 1961; 1962; Murphy, 1962; Kempson *et al.*, 1963 ; Southwood, 1968). These methods are effective against mite species because they move downwards in response to the stimuli of drying soil and heat. Ascid mites are collected in vials with moist POP-Charcoal substrate by using Tullgren funnels (Walter and Lindquist, 1995).

8.5.1 *Litter Sample*

After removing the upper layer, litter sample is collected in a polythene bag. From one tree, 2-3 sub samples are sufficient to make one composite sample. Oribatid mites constitute dominant fauna in litter samples.

8.5.2 Soil Sample

Soil samples are collected from two depths *i.e.* just below the upper layer of soil and soil from the root zone area. Depth of root zone area depends on the age and type of plantation. From field and vegetable crops 10-15 sub samples of 100 cc each are drawn from one-hectare area in a zigzag manner (Figure 8.9). All the sub samples are pooled together in the same polythene bag to form one composite sample. An aluminum foil label showing the sample number is placed inside the bag and tied with a rubber band.

Samples are collected from ten trees randomly from one-hectare area with 4-5 sub samples of 100 cc each from a single tree. The sub

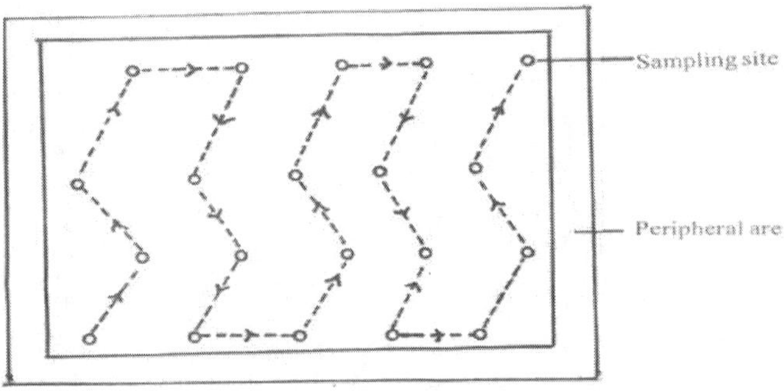

A. Field crops

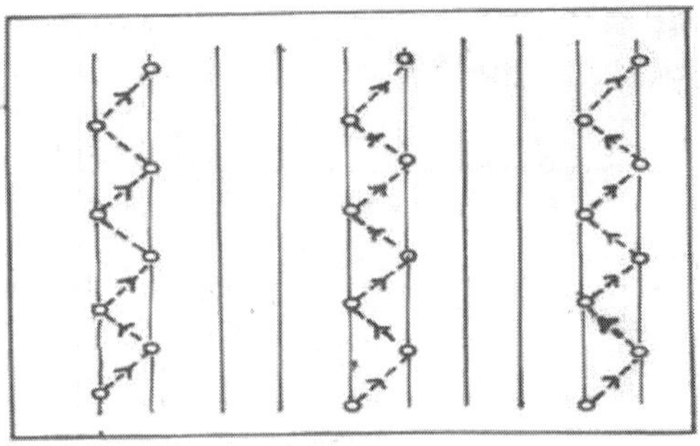

B. Vegetable field

Figure 8.9: Soil Mite Sampling Method

samples are then pooled in a polythene bag containing proper label. In laboratory, each composite sample is mixed well with hands in enamel tray before processing it for extraction. Soil sampling can be done either through cylindrical core or spatula. Depending on the type of soil and nature of animal, different block methods are described by many workers for soil collection.

8.5.3 Hand Sorting

The soil particles are crumbled by hand and spread out in a dish to examine directly under stereozoom microscope.

8.5.4 Sieving

Sieves of different mesh sizes are used to separate mites and other soil fauna. Morris (1927) designed an apparatus for mite extraction. It consisted of three sieves. The first two sieves are floored with perforated zinc with apertures of 3.5 and 1.5 mm diameter. Third sieve placed at bottom had smaller mesh size to retain smallest sized animals and fine particles of soil. Sample placed on upper most sieve is washed through water and residue in each sieve should be examined for fauna. Salt and Hollick (1944) and D'Aguilar *et al.* (1957) also described the extraction techniques based on sieving and floatation methods.

8.5.5 Portable Funnels

Jacot (1932) devised collapsible funnels in which samples are dried either in sun or in a heated room.

8.5.6 Berlese Funnel Method

As originally devised, Berlese's apparatus (1905) consisted of a metal funnel equipped with hot water jacket. Sampling material is kept on a sieve over the wide opening, which slowly dries out due to heat from water jacket. In order to escape the desiccation, mites move down from the sample through sieve and funnel into the collecting tube. Tullgren (1918) replaced the water jacket by an electric bulb suspended over the funnel as a source of heat. In modified Berlese-Tullgren method, the sample is kept on 20 mesh metallic sieve placed on a funnel. At top of funnel, light and heat source is provided through 60W electric bulb for 24h. Due to their photonegative behaviour, mites moves down and are collected in specimen tubes containing water at the end of the base of the funnel (Figure 8.10).

Figure 8.10: Berlese Funnel Method

Glass funnel proves better in mite extraction as compared to metal funnel in which sometimes mites remained trapped due to darkness in funnel tube.

8.5.7 Macfadyen Extractor

In Macfadyen extractor (1955), mites are extracted from soil samples. Samples are taken from the field with stainless steel tubes, the upper ends of which are covered by a metal gauze. Tubes containing samples are inverted and placed in steep sided funnels, heat being derived from a resistance wire suspended above. Murphy (1958) gave 'split funnel' extractor, which is similar in principle (Figure 8.11). For expedition purposes, collapsible plastic funnel sheets may be used. Solomon (1946) gave a field extractor in which the organic material is dried out by heat from a water bath heated by spirit lamp.

8.5.8 Floatation Technique

Ladell (1936) gave this technique for soil mite extraction in which the sample is stirred in a magnesium sulphate solution of high specific gravity and at the same time a stream of air is bubbled into the suspension below. After setting, the liberated animals along with debris float to the surface and are collected on a filter paper or sieve.

8.5.9 Modified Tullgren Funnel

The funnel consists of an aluminium core lined with a cylinder of Plaster of Paris that supports two sieves (Von Torne, 1962). Within

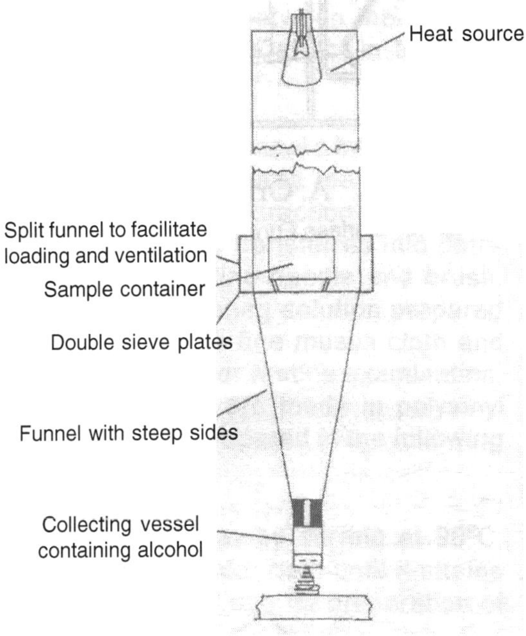

Figure 8.11: Split Funnel Extractor (Murphy, 1958)

the funnel outlet a piece of thick walled rubber tube is inserted, which tightly fitted on the collecting tube. This rubber tube prevents the escape of mites from the funnel outlet. The different funnel methods used for extraction of soil, stored and parasitic mites are also discussed in the chapter.

8.5.10 Open Brass Funnel Apparatus

Haq and Ramani (2003) slightly modified the Berlese funnel method in which at a time thirty samples can be processed for extraction. Collecting vial beneath each funnel rests on a metal-based spring over the iron sheet.

8.6 Storage Mites

8.6.1 Collection of Infested Material

Storage mites are found on unprocessed and processed cereals, pulses, millets and different kinds of nuts. Meat, fish, fish products, dried eggs, mushrooms, cheese, copra, hay, deep litter of broiler houses, disused bee hives, butches offal, stored flower tubers or bulbs also harbor different mite species, which can be collected in polythene bags and subjected to microscopic examination or suitable extraction technique.

8.6.2 Collection of Grains, Residues and Dust

Grain residues and dust of granaries and warehouses provide rich culture of stored mites. Mites prefer flour forms of the grains rather than whole grains. Scarified grains provide them suitable area for their establishment in whole grains. Samples of the flour should be collected from the surface to the depth of few inches whereas from whole grains stored in jute bags, samples should be collected from four vertical strata each consisting of one fourth of grains, including bottom of bag. Many species prefer stored food with a high fat and protein content. Grains stored with water content of 14 per cent or more are more prone to mite attack. Different samples should be collected in different polythene bags, each bearing sample number. Other details about the sample, its locality etc. should be noted down in notebook separately.

8.6.3 Modified Berlese Method

Tullgren (1918) gave a modified version of original Berlese extraction technique described under 8.5.6 subheading, which is most frequently used for mite extraction from storage samples.

8.6.4. Collection of Eggs

Acarus siro eggs are separated from other developmental stages from the culture medium through sieving from 180 µm aperture sieve (Thind and Muggleton, 1998). The sieved eggs then can be used for fresh culture, biology or bioassay studies.

8.7 Water Mites

Samples of red algae, lichens *etc.* are collected in polythene bags. These are placed in 1 mm mesh sieve, which is then washed with filtered seawater twice. Filtrate will carry mites along with other members of meiofauna.

Water mites are collected with a hand net with a mesh-size of 250µm. Different habitats such as riffles, pools, moss carpets, etc. are sampled separately at each site by disturbing and turning them by hand, In this method, all the benthic invertebrates are washed downstream into the net (Goldschmidt, 2002). After being washed through a course sieve (mesh-size 2-3mm), the samples are transferred into a white Frisbee and in most cases revised directly in the field. It is difficult to track a fast swimming water mite in a little jar. Transferring the specimen to a watch hour glass make it easier to handle. A pipette with a wide opening must be used. Halacarids can be collected from sediments in standing and running water, as well as from their crayfish hosts. Habitat characteristics such as water-temperature, pH, conductivity, stream current, stream size, shading, micro-habitats, surrounding vegetation, etc. have been registered.

Sampling Details

Each polythene bag should bear a sample number on an aluminium foil label, which should be put inside the bag. Tie the polythene bag with a rubber band. Note the details of the sample in a notebook mentioning sample number, host plant/grain/animal, date of sampling, name and address of site, symptoms on crops or animals. In case of soil and plant mite, cropping sequence should also be written.

Sample Storage

Samples should be stored in BOD incubator set at 27±1°C and 80-85 per cent RH for few days if immediate processing is not possible.

Chapter 9

Preparations and Mounting

For detailed examination of specimen, microscopic preparation is necessary. Secondly, if extracted mites are to be retained for accurate identification and maintainence of permanent record, they need to be transferred to microslides for mounting. Depending on the degree of sclerotization, suitable media for preservation and mounting is selected.

9.1 Preparation for Binocular Studies

Examination of living mites under binocular microscope is necessary to obtain information on the colour and shape of the body. Methods for preparing the mites for temporary and permanent mounting are given below:

9.1.1 Temporary Mounting

Mounting fluids for temporary mounting are water, glycerine, lactic acid and lactophenol. On a clean glass slide or preferably on a cavity slide, a drop of mounting media is placed on which specimen is released to observe under microscope.

Lower concentration of lactic acid (50 per cent) is used for weakly sclerotized forms and higher concentrations for strongly sclerotized specimen. Slide should be warmed over an electric light bulb. Lactic acid helps in dissolving the soft parts of the body without causing shrinkage of the cuticle. For poorly tanned species, few workers add trace amount lignin pink in lactic acid and glycerin mixture (60:40), which stains the cuticle. Use of lacto-phenol (Phenol 20g, lactic acid 16.5 ml, glycerin 32 ml, distilled water 20 ml) for soft-bodied mites should be avoided as it causes shrinkage in them.

Grandjean (1949) devised carbon block technique for examination under microscope (Figure 9.1). A small surface area of a block of porous carbon is kept moist with 75 per cent alcohol siphoned through capillary tube from a small reservoir. The tube opens on to the block at a particular point where the surrounding region is saturated with alcohol and appears very black as compared to other regions of block.

9.1.2 Killing and Preserving

Majority of the mites except eriophyids are killed and preserved in 70 per cent alcohol with a few drops of glycerine, which do not allow the specimen to dry out. Formaldehyde based preservatives should be avoided as they shrivel the mites.

AGA medium (8 parts 70 per cent isopropyl/ethyl alcohol, part glacial acetic acid, 1 part glycerine, 300 g sorbitol added to one gallon

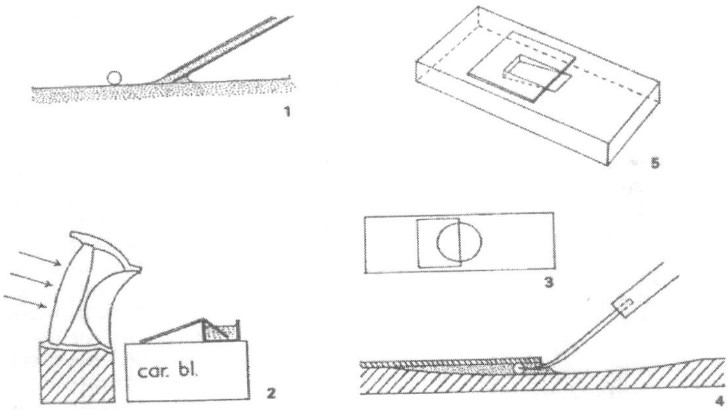

Figure 9.1: Carbon Block Technique (Grandjean, 1949)
By: Incident light (car. Bl., carbon block) (1, 2); Glass (3) and Perspex cavity slides (5); Manipulation of a specimen in a cavity slide (4)

water) is also a good media to preserve the mites as it prevents the mites from hardening. Some workers use boiling water for killing the mites as it causes the extension of the trophic and ambulating appendages, which facilitates the detailed study of these structures.

Preservative for eriophyid mite is thin sorbitol syrup in a 25 per cent solution of isopropyl alcohol. A small amount of iodine crystals should be added in the medium to aid in preservation and to prevent growth of mold.

Oudeman's fluid ethyl alcohol (70 per cent) 87 parts, glycerine 5 parts and glacial acetic acid 8 parts is a better alternative to alcohol for terrestrial mites or soft-bodied species as there is less shrinkage. Viet's fluid (Glycerine 10 parts, acetic acid 3 parts, distilled water 6 parts) is preferred for fresh water mite species.

9.1.3 Clearing

Before mounting the specimen on the slide, it should be cleared completely. Individual mite specimen except oribatids are picked with the help of a pick from the preservative medium and placed on a cavity slide containing a drop of lactic acid solution (50-95 per cent concentration). Gentle heating of slide on the flame of spirit lamp clears the specimen. Process should be continued till the specimen is cleared completely.

For heavily scleratized species, Nesbitt's solution is used. The composition of the medium is: Chloral hydrate (40 g), water (25 ml) and hydrochloric acid (2.5 ml) mixed in the same order. The solution should be kept in tightly stoppered bottle to prevent evaporation and crystallization of chloral hydrate.

For clearing of oribatid mites, transfer them to absolute alcohol and lactic acid (1:1) for one month. During this period, alcohol will evaporate slowly and cleaned specimen can be preserved in pure lactic acid till identification. If oribatid specimen still remains opaque, it should be placed on a cavity slide containing sodium perborate solution prepared in distilled water. Gentle heating of slide will clear the specimen. To hasten this process, aqueous solution of potassium hydroxide (5-10 per cent) may be added to the solution, which causes bleaching of the specimen.

Gentle pressing out of gut contents through a body incision is necessary for blood feeding mites. They are cleared/macerated in lacto phenol (lactic acid 50 parts, phenol crystals 25 parts, distilled water 25 parts) and 5-10 per cent aqueous solution of sodium hydroxide. Digestion in trypsin solution (0.2 g powdered trypsin dissolved in 10 ml of 0.5 per cent sodium carbonate) in an atmosphere saturated with tolerance for 24-72 hours is beneficial for Holacarid mites. Before mounting, they should be first transferred to distilled water.

9.1.4 Staining

After clearing, staining of specimen could be done. Dip the specimen in 1 per cent solution of acid fuchsin or mercuro chrome for one minute and rinse it in water. Iodine and lignin pink can also be used in acarine staining.

9.1.5 Permanent Mounting

Permanent mounts are necessary for building up reference collection and an alternative to temporary mounts for identification. If possible, live mites should be preferred for preparation of microscopic slides over specimen stored too long in preservative. In preserved specimen, sometimes, it becomes difficult to strengthen their appendages, which bear important taxonomic characters.

Cleared specimen is transferred in a drop of Hoyer's medium (50 ml distilled water, 30 g gum arabica, 200 g chloral hydrate, 20 g glycerin-mixed in the same order) on microscopic glass slide, neatly and carefully. Specimen is mounted either dorsally or ventrally in the middle of the drop. Appendages are spread carefully to expose maximum morphological features. Three small pieces of cover slip are placed radially near the specimen to avoid crushing of specimen. With the help of forceps, a clean, small, round cover slip is placed on the mite gently. Air bubbles or air space can be avoided by adding additional medium through the sides of the cover slip.

For oribatid mites, less amount of chloral hydrate is preferred in mounting media, known as de-faure's medium (gum arabica 30 g, chloral hydrate 50-150 g, glycerol 20 g and distilled water 50 ml). PVA medium (Boudreaux and Dosse medium) is also used for making permanent mounts of mites but specimen are more difficult to recover from this medium. Another care that one must take with polyvinyl alcohol (PVA) medium is that slides should not be over heated otherwise shrinkage of medium will occur and cause damage to the specimen.

CM medium (Clark and Morishita, 1950), which contains a high proportion of lactic acid, is also used for permanent preparations. The composition of the medium is: Metho-cellulose 5 g, carbowax 2 g, diethylene glycol 1 ml, 95 per cent ethyl alcohol 25 ml, lactic acid 100 ml, distilled water 75 ml. The metho-cellulose is mixed in alcohol, then added to remaining ingredients and filtered through glass wool. The medium is placed in an oven at 40-45°C for three to five days depending on the requirement of consistency.

The water mites are difficult to study but a stereomicroscope may be very useful for observing water mites. Mites are picked up with tweezers and eyedroppers and fixed in Koenike's solution (glycerine (10 parts): acetic acid (3 parts): water (6 parts)). It is useful to add dots of vaseline under the corners of the cover glass. Since most water mites have a thick spherical body they will otherwise be crushed.

9.1.6 Ringing

After placing cover slip on the mounting specimen, slide is air dried for a week, cleaned with water before ringing with nail enamel/varnish. Ringing is nothing but sealing the outer brim of cover slip to make it more permanent. With the help of brush, nail enamel is applied on three-four points of outer brim of cover slip, which helped in fixing the cover slip. Then, outer brim is sealed thoroughly and completely.

9.1.7 Labeling

Square labels are placed on left and right sides of glass slide. On one side, information should be given about the host, locality, date, collector's name, mounting media and code number to associate it with details written in your notebook. On other side, identity of specimen and identifier is provided.

9.1.8 Recovery of Specimen

Gum chloral preparations are not completely permanent and if necessary the specimen can be removed for remounting by soaking the slides in warm water. Keifer provided a method for recovery of eriophids. Removed mites are placed in Keifer's solution, which consists of resorcinol (50 g), diglycolic acid (20 g), glycerol (25 ml), iodine (enough to produce desired colour) and water (10 cc; enough to produce desired consistency).

After treatment in Keifer's solution, mites are transferred to another solution to eliminate the excess resorcinol. This solution consists of starch free white karo syrup (25 cc), chloral hydrate (125 g), glycerol (5 cc), iodine (moderate amount) and water (15 cc). The recovered and treated specimens are then mounted in Keifer medium (Karo syrup 12 cc; chloral hydrate 60 g; small amount of potassium iodide; iodine crystal 2 g; formaldehyde solution to form thin mixture).

9.2 Preparation for Electron Microscopic Studies

Detailed study of morphological features particularly body seta require electron microscopic studies. Standard methods followed for scanning electron microscope and transmission electron microscope are given below:

9.2.1 SEM Studies

The specimens are fixed in 0.2 M sodium phosphate buffer (pH 7.4) containing 4 per cent glutaraldehyde (EM grade) overnight at 4°C and rinse it three times with buffer. Post fixation is done in 1per cent osmium tetra-oxide in phosphate buffer for two hours at room temperature. Dehydration is through graded ethanol series followed by drying through critical point method. Dried specimen is coated with gold and mounted on specimen stub.

9.2.2 TEM Studies

Before the specimen is fixed for TEM studies, it is necessary to carefully pierce its body with a tiny sharp insect pin. In adult specimen, caudal portion and in nymphs middle part of the body is pierced. Specimen is fixed in 2.5 per cent gluteraldehyde in 0.1 M phosphate buffer (pH 7.4) for 2-4 hours. The specimen is washed several times in pure 0.2 M phosphate buffer or in 6.85 per cent sucrose in 0.1 M phosphate buffer. Then, it is post fixed in 0.1 M phosphate buffer containing 1-2 per cent osmium tetra oxide and 8.56 per cent sucrose overnight. After dehydrating in alcohol acetone series, the specimen is embedded in an araldite mixture. Ultra-thin sections are cut on ultra microtome. For preliminary studies, few sections should be stained in toluidene blue and observe under light optical microscope. For TEM, stains used are uranyl acetate and lead citrate.

9.3 Preparations for Anatomical Studies

Specimens are dissected in the isotonic sucrose solution for anatomical studies and different systems can be examined under phase contrast microscope. For ulra-structure study, electron microscopic preparations should be made as described earlier in the chapter. To study the structure of internal organs, it is considered better to remove the anterior portion of body to allow better penetration of the fixative. Witalinski *et al.* (1990) described a technique for anatomical preparations in which after primary fixation (20 h), the specimens are post fixed with 1.4 per cent osmium tetraoxide in 0.45 M sucrose for 1.5 hour before dehydrating them in graded series of ethanol and propylene oxide and embedded in wax. For lightly sclerotized mites paraffin wax is preferred but for heavily sclerotized mites, ester wax should be used.

Semi-thin sections are stained with Stevenel's blue (Del Cerro *et al.*, 1980) and ultra-thin sections are usually contrasted with uranyl acetate and lead citrate.

9.4 Preparations for Histological Studies of Damaged Material due to Mite Infestation

Histological techniques are described by vairous research workers (Johansen, 1940; Jenson, 1962; Purvis *et al.*, 1964; Rabindra *et al.*, 2003), which involve collection of material, fixation, embedding, sectioning and staining.

9.4.1 Collection of Material

Leaves showing different levels of damage alongwith healthy leaves are cut in 4-5 mm length. Care should be given so that tissues are neither damaged nor crushed.

9.4.2 Fixation

Fixation is necessary to avoid distortion of the material. It helps in preserving the morphology of different structures, stabilizes the chemical constituents and kills the cells to prevent autolysis. Formalin-aceto-alcohol (FAA) is generally used as it penetrates the plant cells. FAA medium consists of 35 per cent formalin (5 parts), glacial acetic acid (5 parts) and 70 per cent ethanol (90 parts). Materials should be immersed in FAA for 24 hours before conserving them in 70 per cent ethanol for further processing.

9.4.3 Dehydration

Dehydration must be done gradually in order to prevent cell distortion. A series of baths are used for this purpose.

Bath 1. Water: 90 per cent Ethanol + Tertiary Butanol (30:50:20). Keep the material for 12 hours.

Bath 2. Water: 90 per cent Ethanol + Tertiary Butanol (15:50:35).

Bath 3. Water: 90 per cent Ethanol + Tertiary Butanol (5:40:55).

Bath 4. 90 per cent Ethanol: Tertiary Butanol (25:75)

Bath 5. Tertiary butanol (100)

Keep the material for 3 hours in each bath from second to fifth bath. After the last bath, materials should be transferred to pure TBA for 12 hours with two changes of this solvent.

9.4.4 Paraffin Infiltration

After dehydration, wax flakes are added to the material in equal proportion. The mixture is left at 20°C for 5 hours, then at 40°C for 12 hours and finally at 60°C for 2 hours in an oven. The leaf bits should be at the bottom of the liquid. At the end of this process, decant the supernatant liquid, add pure liquid and leave at 60°C for two hours. Repeat this last step at least three times.

9.4.5 Embedding

It can be done in small paper vessels on a water bath. Smear the vessels with glycerine and then pour the molten wax carefully at 60°C. Remove the specimen from container with the help of warmed forceps and place them on blocks. If need be, orient them with warm needles. Carry out this process quickly to avoid formation of air bubbles. Leaf bits should be in the middle of block. Cool the blocks in cold water.

9.4.6 Sectioning

Sectioning is done through microtome. The blocks are trimmed with a scalpel so that the embedded sample is visible. Mount it on a small piece of wooden leg. Cover the surface of the object holder with molten wax and allow it to set. With the help of heated spatula, fix the mounted object on holder and orient it in such a way that the sides are in horizontal position in relation to knife. Thin sections of 10-15 µm thicknesses are cut serially, using rotary wheel at room temperature. Sections are placed on a slide coated a *priori* with Haupt's adhesive (glycerine 15 ml, gelatin 1 g, phenol 2 g in 100 ml water) and the constrictions extended with warm 3 per cent formalin.

9.4.7 Deparaffination and Staining

Before staining, wax is removed by xylol and slides are rehydrated. At first, material is passed through xylol and washed for 10 minutes each in a series of rehydration baths in coupling jars in following sequence:

Xylol (2 changes), Xylol+Butanol (50:50), Pure Butanol, Butanol: Ethanol (50:50), Ethanol 96 per cent, Ethanol 80 per cent, Ethanol 70 per cent, and Ethanol 50 per cent.

Staining is done as per the procedure given under the subheading '9.3 preparations for anatomical studies'.

Comprehensive studies on the structure of major internal organs of mites of various habitats are mainly conducted by routine light microscopic methods. In case of acarines, mostly not exceeding 0.5 mm in length, light microscopy alone can create considerable difficulties in interpretation of the observed structures. Investigations done at ultrastructural level could serve as a basis for further experimental approach in development of methods for the control of mite populations.

Chapter 10

Labeling and Housing the Collection

To have any scientific value, collected specimen must be labeled properly giving basic information about it. For slides, square labels one on each side is used. Collector's name, source, date, mounting medium should be written on left label whereas right label is generally reserved for species identification and identifier's name. Maintain a code number system to associate field notes with the specimen collected.

10.1 Housing the Collection

Materials preserved in liquid usually needs no attention other than occasional replacement of preservative and stoppers. Vials should be examined periodically to make sure that specimen does not become dry. Instead of cotton plugs, good quality cork stoppers or plastic screw tops are recommended for vials.

Microscopic slides are usually stored in wooden or plastic boxes/ cabinets. The inner side of the boxes are slotted to hold the slides vertically and to separate them from one another. Before storing slides vertically, they should be thoroughly cured or the cover glass may slip.

If the slides are made with Hoyer's medium, it is desirable to keep the slides horizontally because medium may become soft under very humid conditions, which in turn will damage the collection.

10.2 Packing and Shipping Specimen

Due to their small/microscopic size, there is always a risk of losing or damaging specimen while mailing them. Container should be large enough to include ample packing material to minimize the effects of jarring.

10.2.1 Specimen in Vials

Screw top vials should be used. They should be firmly closed and sealed with plastic adhesive tape before wrapping them separately in cotton, tissue paper or similar material. Care should be taken that no piece of glass comes in contact with another piece of glass. Vials thus wrapped, may be wrapped together or held with tape or rubber bands as a unit and placed in cardboard box with enough packing.

10.2.2 Microscopic Slides

Slides should be thoroughly dried, cured and wrapped in soft tissue paper separately. The slides may be sent in commercially available holders, standard storage boxes or small card board slide boxes with enough soft tissue papers around each end of each slide, between the slides and the box lid to prevent movement. Another alternative is to place thermocoal sheets on all sides of storage box/container. On vertical thermocoal sheets, grooves should be made to fix the tissue wrapped slides vertically. Cotton packing should be given in between two slides, at the margins and inside the lid. If container is of carton material, it should be wrapped in cotton padding and then placed in cloth bag or laminated envelope.

Chapter 11

Culturing Techniques

Rearing of pests whether insects, mites, nematodes or other organisms, has been regarded as non-technical labour in the past. However, due to the enthusiasm, interest and concern of few acarologists working in mid 1900's it led to the establishment of acarine rearing as a developing branch of Acarology in recent years. Mite rearing technology can make direct contributions to the management of many destructive mite pests by enabling strategies that are sound in principle, economically viable and non hazardous to non target organisms. Until appropriate mite number become available, studies on different aspects of mites including genetics of economic pests, plant resistance, biological control are hampered.

An understanding of the mating habits, preoviposition and oviposition period, fecundity, longevity, sex ratio, environmental requirements, food and feeding preferences of mites is necessary in developing rearing techniques. In first section of this book (Acarine biodiversity) these points are discussed in detail to provide an idea about different groups of mites.

Mass rearing of beneficial mites is often a major bottleneck in integrated pest management programmes. Due to slower dispersal rate of many bioagents as compared to pests, producing and delivering large quantities of bioagents is a pre-requisite to any biocontrol programme. Secondly, bioagent mass rearing should provide sufficient number of releases to control pests, their colonization in various ecological zones, allow releases throughout the year and provide material for large-scale experiments. Culturing natural enemies is a complex matter as it involves the rearing of natural enemy as well as the target pest and the multiplication of suitable host plants. Such triple-phased rearing system requires the solution of biological, technical, numerical and logistical problems. The general requirements of mass rearing are provision of pure cultures, free of contaminants, production of healthy individuals with a broad genetic base in a reliable and predictable manner and the development of a time and cost efficient method of production. A flow chart provided by Megevand *et al.* (1993) is given (Figure 11.1) for maintainence and continuous production of phytoseiids in laboratory/greenhouse conditions.

Special and different techniques are necessary for culturing different groups of acari due to vast differences in their habitats. They are too small when compared with insects but larger than other invertebrate fauna like nematodes. Earlier studies on acarine rearing pertained to designing of rearing cages and natural hosts as diets for majority of acarine groups. However, new concepts such as artificial diet, modified environment, quality control are developing. In this chapter, we have tried to gather past and present methods and trends in mite rearing and finally discussing future research plans for young/ new workers in this field.

11.1 Rearing Cages and Observation Arenas

The needs of the rearing programmes, behaviour and habitat of mites decide the designs of rearing cages. Size and shape of rearing cages vary according to the type of food and behaviour of mites. Some desirable characteristics shared by these cages include allowance of proper gas exchange, moisture regulation, airborne contamination, prevention of escape, economics, visibility, accessibility, convenience of handling and harvesting. Mites must have a favourable micro environment before they start multiplying. Size and shape of the rearing cages, durability, visibility, accessibility, and behaviour of mites are some of the other criteria, which should be kept in mind while selecting or designing the rearing cages.

Size and shape of rearing cages vary according to the type of food and behaviour of mite. The cage/container must hold the mite close to the diet and minimize the wandering (Ranjana *et al.*, 2005). In addition to access to diet, shape of the cage also influences the

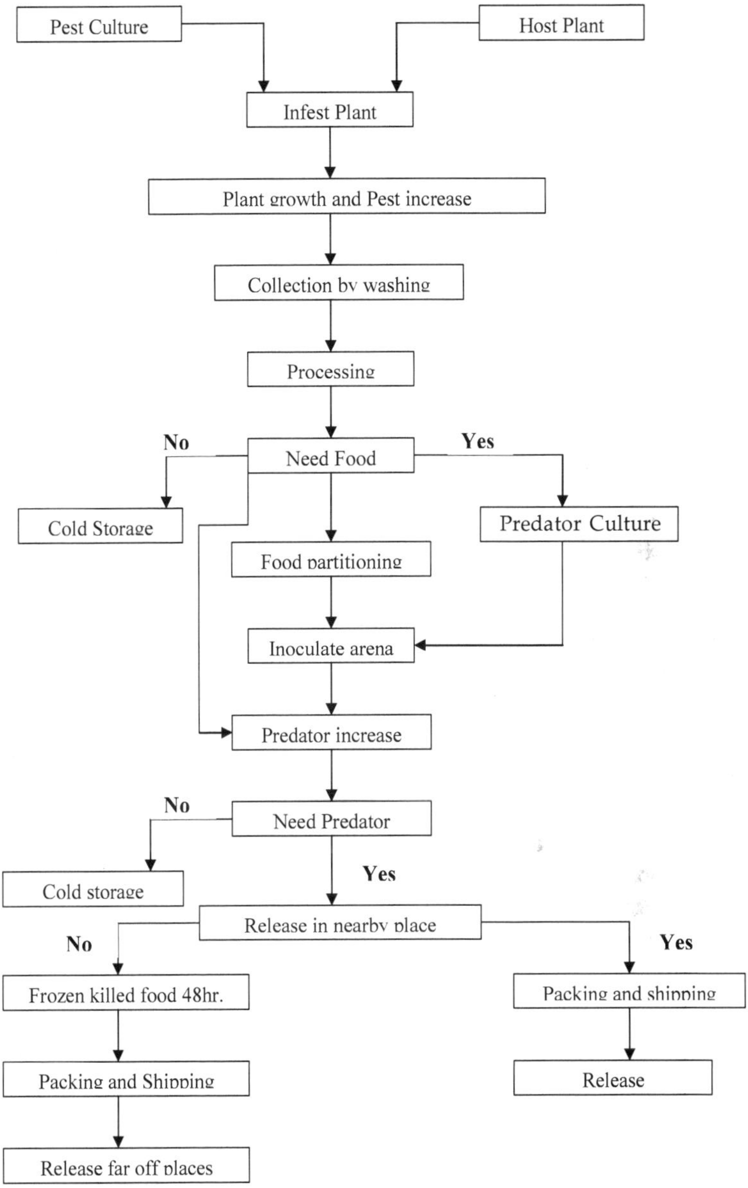

Figure 11.1: Flow Chart for Phytoseiid Maintainence and Production System (Megevand *et al.*, 1993)

moisture retention, which regulate the drying rate of the diet. Rearing cages/observation arenas are normally placed in desiccators containing potassium chloride/potassium hydroxide/zinc sulphate solution, particularly used in case of stored, soil or house dust mite cultures. Population density, which depends partly on rearing cage size, can also affect the growth and fecundity of mite. High population density per unit area causes shortage of food and accumulation of mite excreta, which would generate minty odour, excessive moisture in cage, promote mold growth and hot spot formation in cage. Subculturing and food replacement are also important in maintaining the culture of particular species.

In the case of reusable containers, durability is important in reducing the replacement costs. Containers that withstand the sterilization temperature are preferred. Containers should be resistant to appropriate diets in case chemicals are, being used as a component of diet. Frequent visual inspection is necessary to maintain the culture and its quality. Both visibility and accessability are also important which save valuable time when mites are removed from cages at various growth stages.

Details of different rearing cages used by different acarologists are given below, which vary with different groups occupying different habitats in an ecosystem.

11.1.1 House Dust Mites

No special cages are required for house dust mite cultures. Rearing cages consists of plastic petri dishes with tight fitting lid, as described under '11.1.6 stored mites' are used for mite rearing. Natural food or artificial diet as mentioned later in this chapter could be used for multiplication. As these mites are encountered in humid conditions, which remain unexposed to sun, faster development could be achieved by placing these cages in desiccators containing super saturated solution of sodium chloride/potassium chloride. These should be kept at 25°C in BOD's. Winston and Bates (1960) demonstrated that 0, 33 and 93 per cent relative humidity could be obtained by using calcium sulphate, magnesium chloride and potassium nitrate, respectively.

11.1.2 Parasitic Mites

Arlian *et al.* (1989) used cylindrical glass cages (17×6m) as observation arenas for *Sarcoptes scabeii*. In these cages, both the ends are sealed by nylon mesh with the help of plugs and are placed in wire mesh baskets, which is suspended over glycerol solution in 200 ml glass jar. Glass jar is covered with rubber stoppers. Aqueous glycerol solution in jar provided the desired relative humidity as described by Segur (1953).

The culture of *Dermatophagoides gallinae* (Chicken mite) is raised in plastic containers with a fine mesh screen lid and lined on the inner side with corrugated cardboard (Fletcher and Axtell, 1991). Suitable food is provided inside the container and whenever required, mites are aspirated into a collection container from the top of the corrugated cardboard lining.

11.1.3 Phytophagous Mites

Cultures are routinely reared on circular leaf discs (2.5 cm diameter) (under side up) placed on moist absorbant cotton in petri plate (popularly known as leaf disc method; Figure 11.2). Placing ordinary filter paper under cotton is helpful in soaking extra water. A layer of cotton is placed around the leaf disc to prevent the escape of mites. Petriplate is properly covered and kept at 27±1°C and 75-80 per cent relative humidity. Leaf discs are changed as and when required. In place of absorbant cotton, many workers use a thick piece of foam. In this case, rearing unit consists of an island carrying blotting paper and thick foam in the centre of a tray, surrounded by water.

Slight modification in this technique was done by Walde *et al.* (1995) who also used petri dishes as rearing units in which leaf disc with underside up is placed on filter paper. Absorbant cotton wool saturated with distilled water containing fungicide acted as substrate for filter paper. These petridishes are suspended in closed clear plastic containers containing saturated sodium chloride solution, which provided required humidity (Winston and Bates, 1960). Dicke *et al.* (1988) used Munger cell as a rearing unit.

Fotedar (1981) raised the sem seedlings in sand filled plastic pots for spider mite culture. Each pot is covered with chimney globe, closed at its mouth with muslin cloth. Lee *et al.* (1990) used sand filled nursery boxes for spider mite cultures on soybean seedlings. Plastic bags (30 cm wide opening) with approximately 5 cm growing medium (soil: peat moss: sand: pumice; 1:1:1:2 parts) at the bottom were used by Zhang and Croft (1995) to culture *Tetranychus urticae* on lima bean plants and maintained in growth chambers at 25-27°C.

Circular discs (1-2 cm diameter) of young tender leaves are also used for eriophyid culture. These leaves are pinned on to the wax bottom in petri dishes (5 cm diameter) to enable floating on water poured to two third of the petri dish. Leaves remained fresh for five to six days.

11.1.4 Plant Predatory Mites

Arenas of leaves placed on saturated cotton wool in plastic petri dishes or glass plates, as described for phytophagous mites, are also used to culture predatory mites (James, 1989; Thistlewood *et al.*, 1992; Walde *et al.*, 1995). Peach leaf arenas (1.5×1.5 cm) for *Amblyseius victoriensis* (James, 1989) and okra leaf arenas infested with *T. cinnabarinus* for *A. potentiallae* are placed on saturated cotton wool in plastic petri dishes. For *A. fallacis* rearing, detached bean leaves infested with *T. urticae* (Thistlewood *et al.*, 1992) placed on water soaked sponge with in a pie pan are used, surrounded by water inside a plastic drawer unit. Leaves are changed as and when required in all the cases. Iation *et al.* (1995) described special rearing units, which consisted of a light weight metal spring clip to which two rings of foam are attached. In between the foam rings, the leaf (15 cm^2) is placed and the end of the cage is closed with a circle of wire gauze and transparent plastic sheet. (Figure 11.3).

Black paper or tile is also used for predatory mite cultures, which prevents the need for changing the leaf often. McMurtry and Seriven (1966) reared phytoseiids on black paper (12×12 cm) resting on water saturated plastic foam in a stainless steel pan (20×20 cm) with *T. urticae* as food. Instead of paper, black plastic plate (10×15 cm) served as substrate for phytoseiids in a technique described by Overmeer (1981). After surrounding the tile with wet paper tissue, it is placed on a piece of foam rubber contained in a small tiny vessel with water. The paper tissue covered about one cm. of the upper surface of the plate along with the four edges and the rest was dipped in water. A barrier of glue is made on the paper tissue along the edges of the tile. Small two sq. cm. roof shaped pieces of transparent plastic are placed as a shelter in order to allow mite oviposition. Pollen diet is provided as food. Morewood (1992) used floating arenas consisting of plastic bingo chips anchored with short pieces of thread in 150

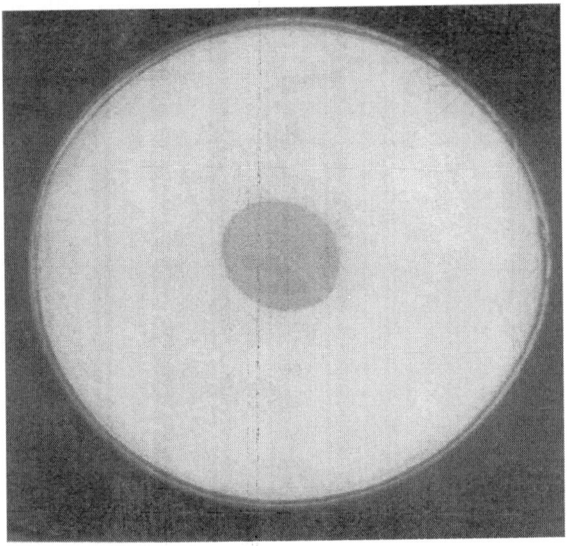

Figure 11.2: Leaf Disc Method

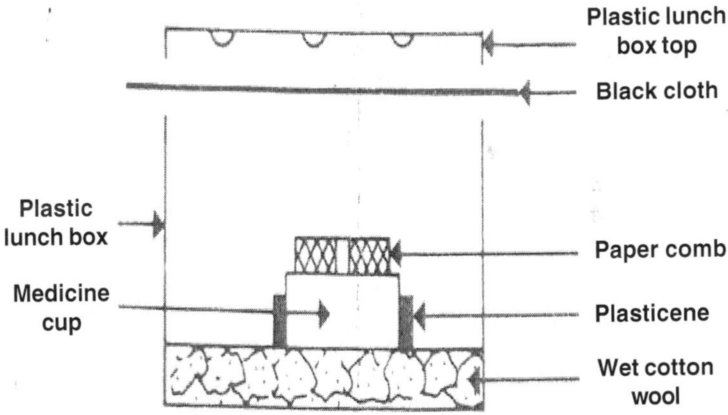

Figure 11.3: Phytoseiid Culture Method (Iation *et al.*, 1995)

mm petri dish containing water for *P. persimilis* culture. Natural host is provided as food.

Technique described by McMurtry and Seriven (1966) was also used by Zhang and Croft (1995) to rear four phytoseiid species and maintained them at 25±5°C, 16L:8D photoperiod and 40-95 per cent relative humidity. For harvesting phytoseiid eggs, they used 50 sq. cm. tile squares surrounded by water saturated cellu cotton in a plastic box (11×11 cm) as rearing arena. The water barrier prevented the mites from escaping. *T. urticae* as food and cotton fibres/tufts for oviposition are provided under cover glass slides. The tile and plastic box are placed in plastic container (30×20×8 cm) with a saturated solution of sodium chloride, which provided required humidity (~80 per cent RH). Cotton tufts enhanced the oviposition rate as Kostiainen and Hoy (1994) found eight fold increase in phytoseiid population. *A (Neoseiulus) fallacis* is also successfully reared in green house on lima bean plants infested with *T. urticae* (Strong and Croft, 1993).

11.1.5 Soil Mites

The first record of culturing mesostigmatid mites is from Michael (1881) when he used small, round dissecting troughs as rearing cells, each covered with a flat plate of glass pierced with two or three holes. The holes are covered with fine muslin, which moved beyond the boundary of holes to prevent the loss of mites and moisture. Suitable food is provided in cells. Sengbusch (1954) used ordinary cover slips in place of fine muslin to cover the cells, kept in position by means of a thin film of vaseline. He placed these cells in desiccators containing saturated solution of zinc sulphate, which provided 82 per cent relative humidity. Maintenance of correct moisture condition was found difficult in this case and drops of water have to be added in the rearing cells at frequent intervals. Searls (1928) achieved a breakthrough in eliminating this difficulty, when he used culture cells with plaster-charcoal or plaster-soil substrate. These cells were further modified by Edwards (1955) and Rohde (1956).

Edwards (1955) used two parts of plaster of paris and one part of fine organic soil to make culture cells. Fine organic soil can be obtained by mixing three parts of soil with one part of dried manure or leaf mould (Evans *et al.*, 1961). Rohde (1956) further modified this technique by using thick paste of nine parts of plaster of paris (POP) and one part of charcoal made in water, a technique still being popularly used in preparing rearing cells or observation blocks for soil mites, which need high humid conditions (Figure 11.4). For this, a thick flowable paste of POP and Charcoal (9:1) is prepared by adding water. A block is prepared in a cavity formed by joining two L-shaped blocks. Closed end of a small test tube (1 cm dia. Or appropriate thickness) is embedded 0.5 cm deep in the middle of block and it is allowed to dry in this position. Test tube should be taken out carefully imparting smoothness to the cavity from inside. Upper surface of the block must be smoothened by rubbing gently with water on a smooth surface. After releasing mites and food, it should be covered with glass slide with the help of rubber bands at the two ends (Figure 11.4; Walia, 1992). These arenas should be kept moist by wetting with water from lower side as and when required. It can be poured into jars/trays depending upon the type of the cell required. This mixture can prepare a convenient series of small cells.

Murphy and Doncaster (1961) used sintered glass filter crucibles as culture chambers. Al Amidi *et al.* (1991) used plastic cups as

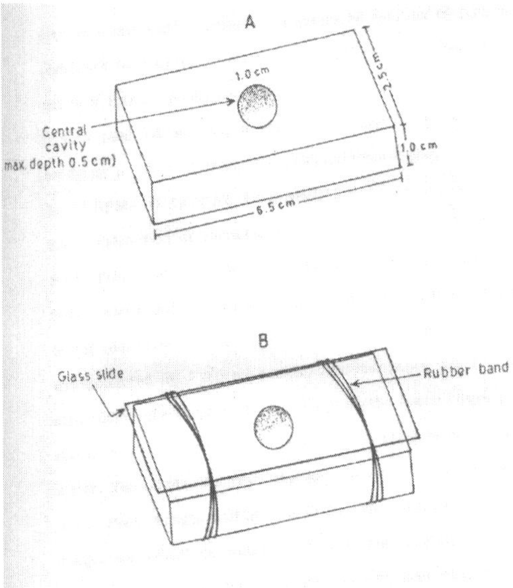

Figure 11.4: Observation Blocks (Walia, 1992)

rearing cells for *Parasitus bituberosus*, a biocontrol agent of dipteran flies in mushroom.

Plastic petri dishes with bottom covered with a mixture of plaster of paris and charcoal (9:1) are also used for prostigmatid and mesostigmatid mites. Trombidid mites whose larvae are parasitic on insects and nymphs and adults are predatory in nature, were successfuly reared by Zhang and Xin (1989). Parasitic larvae are reared in cotton plugged small vials of 3 cm diameter and 1 cm height with suitable hosts. For adults and nymphs, rearing cell of 2 cm diameter and 10 cm height are used with moist soil as substrate. Zhang (1991) modified this technique slightly and used moist cotton balls in place of moist soil in vials with aphids as food. With the help of fine brush, egg masses are transferred to another vial containing 2 per cent agar at the bottom since high humidity is required for egg hatching.

Oribatid mite cultures can be reared in plastic cells/boxes/plates with suitable amount of litter. After covering with lid (which had small opening covered with muslin cloth) or muslin cloth, these rearing cells are placed in BOD at 27±1°C and 85-90 per cent RH.

Rearing cage consisted of plastic pots (13 cm dia.) fitted with a clear plastic top to contain mites is used for *Halotydeus destructor* which is considered a major pest of pasture legumes in South Australia. Pots are filled with sand: loam mix (4:1), fertilizer and bacterial rhizobium mixture to promote root nodulation in sub clover seeds. Mites are inoculated when seedlings are 3-4 weeks in age. Each plastic top (11 cm high) contained stainless steel circular mesh in all the four sides and the lid to increase the ventilation. Pots are maintained in growth cabinet at 4°C, 10 L: 14L photoperiod and 78-96 per cent relative humidity.

Macrochelid mites are cultured in plastic box with 50 mm moist sand. Fresh cow dung layer is placed on the sand and suitable amount of food (eggs of Dipteran flies) are released on cow dung surface. After mite release, cages are closed with glass slide held in place by a metal clip.

11.1.6 Stored Mites

Attempts to culture these mites dates back to 1901 when Michael prepared small rearing units of glass cell. Each glass cell is composed of a 3"×1" microscope slide with a glass ring, 3/8" high and fixed to its centre with balsam. Central ring of 3/4–7/8" diameter is made from glass tubing and covered with a thin piece of glass, which is kept in position by means of a spring clip. The floor of the cell is covered with a piece of moistured filter paper and suitable food material is laid on them.

Solomon and Cunnington (1964) describd Robertson's micro cells for stored mites culturing in which each cell is made from perspex with cell cavity of 5 mm in diameter. They are closed below with black filter paper glued to the base of the cell and above with a coverslip held in place by clips. The cells containing food and mites are placed in glass desiccators containing potassium hydroxide solution, which provided 90 per cent relative humidity.

The plastic petri dishes with tight fitting lids also served as rearing cage for acarid mites (Figure 11.5). A centre hole is created in a lid with the help of red-hot needle, which is then covered with fine mesh silk cloth. This is necessary to provide humidity and air circulation. In these culture arenas, mites are released on suitable amount of food. These culture arenas are kept in desiccators

Figure 11.5: Rearing Cage for Acarid Mites

containing super saturated solution of potassium chloride, which in turn are placed in BOD at 27±1°C and 80-85 per cent relative humidity These mites can be reared on molded bread, yeast or on grains from which they are extracted.

The prostigmatid mites, which feed on astigmatid mites, can be reared in the rearing cages described for acarid mites. Modified rearing cages, consisted of plastic petri dishes whose bottom is covered with a layer of Plaster of Paris and Charcoal (9:1) or observation arenas consisted of glass slides with layer of POP and Charcoal on all sides (Figure 11.6). In the middle, a round hole (½″ dia.) is created to confine these mites with suitable amount of food. The brim of the cell is smeared with a barrier made of Canada balsam mixed with castor and citronella oils.

11.2 Maintenance of Mites on Natural Hosts

The mite rearing on natural host/diet is relatively an easy process for house dust, oribatid and storage mites but considered as a tedious process for phytophagous, parasitic and predatory mites as it involves two or three phase system of rearing *viz.* maintainence of suitable plant culture, host/ rearing organism and/ or bioagent.

Figure 11.6: Observation Arenas (Singh, 1985)

Initial efforts to culture mites of different habitat involve host commodity for majority of pest (stored, house dust mites) and some useful (oribatid and nematophagous) mites. Natural hosts remain a favourable choice till date in mite rearing. Nevertheless, literature pertaining to culture parasitic and predatory mites on atlernate diets has also become voluminous in the last decade.

11.2.1 House Dust Mites

House dust mites are maintained on human skin scales (Van Bronswijk, 1972) and beard shavings (Taylor, 1975) although addition of supplement food in form of yeast induced favourable response. Hart and LeMerdy (1988) used insect meal and beard shavings to raise *Euroglyphus maynei* culture. Use of pork fat in diet also had positive effect on house dust mite rearing (Castagnoli *et al.*, 1989).

Anderson (1988) used powdered yeast and human skin scales (1:1) for culturing house dust mites at 93 per cent relative humidity (KNO_3; Winston and Bates, 1960), 14L: 10D and 22-24°C.

11.2.2 Parasitic Mites

Parasitic mites associated with human beings, other mammals, birds are harmful in nature as they cause some ailment in the host due to their infestation. On the other hand, parasitic mites associated with insect pests are beneficial in nature as they are used as biocontrol agent of a particular host. Amongst the harmful parasitic mites associated with mammals, *S. scabei* is reported to feed on skin crusts of rabbits and dogs (Arlian *et al.*, 1989). Cultures of fowl mite, *O. sylviarum* are maintained on white leghorn hens by transfer of pellets every three months. Whenever required, adult mites are collected by removing infested feathers from birds placing them into an enamel pan and aspirating the dispersing mites into a holding container (Fletcher and Axtell, 1991). Chicken mite, *D. gallinae* culture is raised on blood meals in form of one-week-old chick, provided thrice a week (Fletcher and Axtell, 1991).

Mites belonging to pterygosomatidae family are associated with blood sucking bugs and cockroaches. *Pimeliaphilus plumifer* cultures are reared on Triatomine bugs at 30°C and 15-20 per cent relative humidity (Anderson, 1968). Larvae of *Arrenunus* species (Arrenuridae) are found to be the most common parasite of mosquitoes (*Anopheles, Aedes* and *Culex* species) (Stechmann, 1978). Its nymphs and adults feed on mosquito eggs and immature stages. Mites of *Locustacarus* species (endoparasites) and *Podapotipoides* species (ectopasrasite) are associated with acridids (Gauchat, 1972).

Members of Hemisarcoptidae (Gerson and Smiley, 1990), Pyemotidae (Mallea *et al.*, 1983), Tydeidae and Erythraeidae (Tandon and Lal, 1976) are associated with scale insects. Out of these, methods for mass rearing of *Hemisarcoptes coccophagus* are available. Cultures are reared on armoured scale insects maintained on potato tubers at 80 per cent relative humidity. Initial release of nymphal stages of mites on incised elytra of coccinellid beetle, *Chilocorus bipustulatus* placed on tubers (Gerson and Schneider, 1981; Gerson and Smiley, 1990) starts the rearing process.

11.2.3 Plant Mites

Tetranychid cultures are best maintained on leaves of different bean crops (Scriven and McMurtry, 1971; Fotedar, 1981; Samsoe-Petersen *et al.*, 1989; Lee *et al.*, 1990; Thistlewood *et al.*, 1992). Cultures of *T. cinnabarinus* on okra leaves, *T. neocaledonicus* on okra and cucurbits leaves, *T. ludeni* on cowpea leaves, *T. macfarlanei* on cotton leaves, *Eutetranychus mentalis* and *Panonychus citri* on citrus leaves, *Eotetranychus hirsti* on fig leaves, *Oligonychus indicus* and *Schizotetranychus andropogoni* on sugarcane leaves, *O. coffee* on tea/ coffee leaves and *O. mangiferus* on mango leaves can be reared in laboratory or greenhouse conditions. Among the eriophyids, cultures of *Aceria cajani* on pigeonpea leaves (Janarthan *et al.*, 1972), *A. jasmini* on jasmine leaves, and wheat curl mite on wheat can be maintained in glass/green house conditions (Harvey *et al.*, 1995). *A. litchi, A. mangiferae* and *Acaphylla theae* are considered as important pests of litchi, mango and tea, respectively, and their cultures can be maintained on their respective hosts. Few species of tenuipalpid mites (false spider mites) have been recorded as economic pests of fruit crops and their cultures can be maintained on their natural hosts. *Raoiella indica* (tenuipolpid) and *Eriophyes guerreronis* are major pests of coconut in southern states of India and are maintained on leaves and nuts of coconut, respectively.

When food is exhausted, mites can be transferred to fresh leaves by leaf washing machine (Scriven and McMurtry, 1971), placing infested leaves on fresh leaves/plants (Samsoe-Peterson *et al.*, 1989), using a single hair bristle in a needle holder (Reed *et al.*, 1964) or using aspirator.

11.2.4 Plant Predatory Mites

Phytoseiid predators are capable of using many types of mites including tetranychids, eriophyids, tarsonemids, tydeids, tenuipalpids as well as variety of insects (thrips, white flies, mealy bugs, scale crawlers (Gulati, 2003). Several studies are reported on phytoseiid mites reared on tetranychids and eriophyid mites (Dosse, 1961; Mc Murtry *et al.*, 1970; Van de Vrie, 1973; Amano and Chant, 1977,1978; Tanigoshi, 1982; Helle and Sabelis, 1985; Dicke, 1988; Dicke *et al.*, 1988). Phytoseiid pre oviposition and oviposition are also influenced by the type, size and stage of the prey (Croft, 1972; Croft and McMurty, 1972; Sabelis, 1985; Bruce-Olivier and Hoy, 1990; Duso and Camporese, 1991).

The feeding preference of predators may arise from predation history (learning) or a response to certain environmental conditions (Walde *et al.*, 1995). In various studies, preferences have been linked to energy returns, specific nutrient requirements and prey density (Pyke *et al.*, 1977; Krebs, 1978).

The consumption of eggs and larvae of pest mites favours a high conversion of food into egg biomass thereby determining higher intrinsic rate of increase as compared to feeding on pest nymphs and adults. However, when only prey (*Brevipalpus phoenicis*) eggs are given to *Amblyseius finlandicus* the progeny consisted of all females, which failed to reproduce in absence of males. Oviposition is found to be minimum when only *B. phoenicis* larvae are fed (Sharma and Sadana, 1983; 1989). Gut fullness also influences prey selection for acarine predators like *Phytoseiulus persimilis* (Sabelis, 1990), *Typhlodromus pyri* (Dicke *et al.*, 1988), *Zetzellia mali* (Santos, 1976a; b).

Hoy *et al.* (1982) described a method for large scale rearing of phytoseiid in the field and Overmeer (1985) reviewed various rearing methods of phytoseiidae. *P. persimilis* is commercially mass reared and used as biocontrol agents against tetranychids especially *T. urticae* (Hamamura *et al.*, 1978; Van Lentern and Woets, 1988; Rasmy *et al.*, 1991; Morewood, 1992). Mass rearing techniques have also been developed by King and Morrison (1984) and Krishnamoorty (1988). Rasmy *et al.* (1991) tried to rear this mite on tenuipalpid and eriophyids but concluded them as not suitable prey. *T. pyri* cultures are maintained on *T. urticae* (Bowie *et al.*, 1999), *P. ulmi* (Duso and Camporese, 1991). Duso and Camporese (1991) in the same study successfully reared this mite on *Eotetranychus carpini* and *Colomerus vitis* and reported that an increase in spider mite supply favoured a shorter pre oviposition period and higher oviposition rates. *Amblyseius alstoniae* is a voracious feeder of *Eutetranychus alstoniae* (Dhooria, 1980), *T. macfarlanei* (Jose *et al.*, 1989) and *B. phoenicis* (Kumari and Sadana, 1995a; b). *A.* (*Neoseiulus*) *fallacis* cultures are maintained on *T. urticae* (Thistlewood, 1991; Thistlewood *et al.*, 1992; Coop and Croft, 1995), respectively. *A.* (*N.*) *fallacis* is giving sufficient biological control against *T. urticae* on crops such as hop, peppermint, strawberry and corn (Hadam *et al.*, 1986; Strong and Croft, 1993). Use of acaricide was curtailed in these crops after successful release and colonization of predatory mite. *A. andersoni* cultures are maintained by Duso and Camporese (1991) on *P. ulmi*, *E. carpini* and *C. vitis*. Two other *Amblyseius* species are reared on acarid mites cultured on wheat bran. *A. ovalis* feeds voraciously on tetranychids (Gupta, 2003), whitefly (Borah and Rai, 1989), *Polyphagotarsonemus latus* (Hariyappa and Kulkarni, 1989). *A. longispinosus* is an efficient predator of *T. cinnabarinus* (Pande and Majumdar, 1990) and *T. ludeni* eggs (Mallik and Channabasavanna, 1983). Mass rearing techniques have been developed by Anil *et al.* (1995), Manjunatha *et al.* (1995) and Mallik *et al.* (1999). According to Mallik *et al.* (1999) if 250 predators are released on a batch of fifteen plants, 4700 predatory mites can be harvested per batch per month. *A. tetranychivorous* readily feeds on *T. ludeni* (Puttaswamy, 1978) and *R. indica* (Jagdish and Nageschandra, 1979; 1982) whose mass rearing techniques have been developed by Krishnamoorthy (1990). Another *Amblyseius* species, *A. victoriensis* is reared on *T. urticae* (tetranychid), *Aculus cornutus* (peach silver mite) and *C. vitis* (grape leaf blister mite). The latter two are eriophyid mites, which provide good survival (100 per cent) and oviposition (1.2 eggs/day) (James, 1989). Fungivorous, non-pestiferous tarsonemids may serve as supplementary diets for some predaceons phytoseiid mites (McMurtry *et al.*, 1970). Stigmaidae is another predatory family associated with tetranychidae and eriophyidae. Rasmy *et al.* (1987) used an acarid diet to rear *Agistemus exsertus* for use in spider mite control. *A. fleschmeri* feeds on *B. obovatus* (Chatterjee and Gupta, 1996), tetranychids (Gupta and Chatterjee, 1997) and *A. mangiferae* (Gupta and Gupta, 1992). Biology of *A.*

herbarius on *O. mangiferus*, *T. urticae* (Rather, 1999) and *A. industani* on *T. ludeni* (Rai and Singh, 1999) is studied in detail and cultures are reared. *A. terminalis* is recorded to feed on *Brevipalpus* species and *Acaphylla theae* (Gupta, 2003). *Zetzellia mali* cultures are reared on *P. ulmi* and *Aculus schtechtendali* (Santos, 1976a, 1991; Clements and Haormsen, 1993). *Z. mali* has a low consumption rate (Santos, 1976b) and low mobility (Lawson and Walde, 1993) relative to phytoseiids but has a high reproductive rate (Santos, 1976b).

Mites belonging to Anystidae family are also found to be associated with tetranychids (Rishi, 1990; Gupta and Gupta, 1992). *A. agilis* is found to be an important predator of grape leafhopper, *Erythroneura elegantula* and wooly pine needle aphid, *Schizolachnus piniradiata*. Female mite consumes 5-6 leafhopper nymphs per day and lays 33 eggs in her lifetime. Two generations are raised per year (Sorensen *et al.*, 1976). Long generation time and cannibalistic habits precluded it from consideration as a candidate for mass rearing in laboratory.

11.2.5 Soil Mites

Majority of the soil mesostigmatid mites are able to survive, reproduce and culture on nematodes (Walter and Ikonen, 1989). For this, Potato dextrose agar (4:1) was prepared and poured in 250 ml capacity flasks. After sterilization, flasks should be allowed to cool in a tilted position to get the slants. These slants should be inoculated with small piece of fungus or fungus with fungivorous nematode. After a week, 2-3 female mites should be released which will multiply and form a rich culture.

Potato dextrose agar (PDA) medium can be prepared by following method: Peel, chop and boil the potatoes (200g) in 500 ml distilled water and filter through muslin cloth. Mix the filtrate (potato extract) in already boiled 20g agar in 500 ml distilled water. Add 20g dextrose in this mixture. Pour 150 ml of this medium in each Erlenmeyer flask (250 ml) and steam sterilize at 1.0546 kg per cm^2 for 20 minutes. Keep the flasks in tilted position till hot and store in refrigeration for further use.

Macrochelids are routinely mass reared on a medium containing eggs and immatures of house flies (Cicolani, 1979; Wallace *et al.*, 1979 ; Anderson, 1983) to which nematodes are added (Willis and Axtell, 1968). Another predator of the house fly larvae, the uropodid *Fuscuropoda vegetans* is cultured on the same diet. It lays maximum eggs on nematode diet on which it lays seven eggs per day followed by an house fly immature diet (four eggs/day) (Jalil and Rodriguez, 1970a; b).

Some members of the family Laelapidae are soil borne. *Androlaelaps* and *Strateolaelaps* spp. are fed on corn rootworm, *Diabrotica longicornis* and *D. virgifera* eggs in the laboratory. A generation is completed in about two months at 20°C and 95 per cent relative humidity (Mihm and Chiang, 1976). Another soil-borne predator, *Hypoaspis aculeifer* is reared on plant nematode, *Tylenchorhynchus dubius* in pots (Sharma, 1971; Van de Bund, 1972) and on acarid mite *Caloglyphus* sp. (Lesna *et al.*, 1995). *H. miles* cultures on two acarid mites, *Rhizoglyphus echinopus* and *Tyrophagus putrescentiae* are maintained in the laboratory (Lesna *et al.*, 1995). Walia (1993) reared *H. calcuttensis* on plant parasitic nematodes in the laboratory and in pots placed in screen house.

Among the soil-borne ascid predators, *Arctoseius cetratus* is a voracious feeder of a sciarid fly, *Lycorella auripila*, a pest of cultivated mushrooms (Binns, 1973). *Platyseius* and *Cheiroseius* spp. are found to feed on mosquito eggs and larvae (Smith, 1983) hence cultures can be reared on this prey. Different *Lasioseius* species are cultured on nematodes like *Aphelenchus avenae* (Imbriani and Mankau, 1983), *Anguina tritici, Meloidogyne incognita* and *Longidorus pisi* (Ranjana, 2002). Walter (1987) cultured *Gamasellodes vermivorax* on nematodes. *Lasioseius* species is also successfully cultured on fungivorous nematode, *Aphelenchoides composticola* in potato dextrose agar (PDA) medium (Ranjana, 2002). For preparation of PDA medium, potatoes (200g) are peeled, chopped and boiled in 500 ml distilled water and filtered through muslin cloth. Filtrate is then mixed with already boiled agar (20g) in 500 ml distilled water and dextrose (20g). Flasks containing PDA medium are steam sterilized, kept in tilted position till hot and stored in refrigeration for further use. Each flask is inoculated with a small bit of *A. composticola* and *Fusarium solani* culture. After seven days, few adult mites are released in the medium, which survived, reproduced and multiplied successfully.

Poecilochirus monospinosus (Parasitidae) is cultured on housfly immatures in poultry manure (Wise *et al.*, 1988). Another member of the same family, *Parasitus bituberosus* is reared on mushroom compost containing a dense suspension of free living nematodes (Al-Amiidi *et al.*, 1991). *Eutromibidium locustarum* (Trombidiidae) produced a generation in sixty days in laboratory when reared on grasshoppers (Severin, 1944). Cultures of *Allothrombium pulvinum* are maintained on aphid colonies on bean leaves (Zhang and Xin, 1989). Welbourn (1983) listed several trombidiid mites which have a high potential for the biological control of insect pests and may be mass reared in the laboratory for field releases.

11.2.6 Storage Mites

One of the most important contributions in rearing stored mites was the use of wheat germ as a primary nutrient source by Soloman (1946). He used wheat germ for *Acarus siro* rearing which became the basis for rearing many other stored mite species *viz. Suidasia nesbitti* (Singh, 1985), *Tyrophagus putrescentieae* (Kohli, 1993).

T. putrescentiae cultures are also reared on scarified grains/flour of wheat, redgram, pearlmillet, groundnut (Kohli and Mathur, 1994), cheese (Yoshizawa *et al.*, 1971) and on various nematodes (Walter *et al.*, 1986; Walia, 1993; Ranjana, 2002) at 27±1°C and 80-85 per cent relative humidity. Cultures are also maintained on dried mushrooms at 22°C and 100 per cent relative humidity (Krezeczkowski, 1961) and on bread (Rodinovsky and Krantz, 1961). However, coffee beans, cocoa beans and few legumes (pea, beans, lentil) are not found suitable for its development.

S. nesbitti cultures are also maintained on scarified grains of wheat and pearlmillet (Singh, 1990) and pulses (Singh and Gulati, 2001) whereas *S. medenensis* preferred salted sun dried fish but found to be sensitive to salt concentration (Cicilykutty *et al.*, 1981). *A. siro* cultures are also maintained on chedder cheese (Peace, 1983). *Lardoglyphus konoi* is mass reared on fish powder but *L. kramari* preferred cereals and bread (Rodinovsky and Krantz, 1961). Addition of dried yeast powder in the diet showed its attractiveness and suitability to development of many acarid mites (Maurya *et al.*, 1983). Wheat germ/flour: yeast (4:1) is a standarized diet to rear acarid mites (Solomon, 1946; Parkinson *et al.*, 1991; Kohli, 1993).

A. siro cultures can be reared on wheat germ and yeast in the ratio of 3:1 (Solomon, 1951) at 15±0.5°C and 90 per cent relative humidity in constant dark environment (Thind and Muggleton, 1998).

Among the parasitic mites associated with stored insect pests, mass culturing techniques of *Pyemotes tritici* (Rangel, 1901; Weiser, 1963; Bruce, 1983, 1989), *Blattisocius tarsalis* (Haines, 1981) and *Acarophenax lacunatus* (Faroni *et al.*, 2001) on natural hosts are known. Rangel (1901) was the first to rear a beneficial mite in laboratory, namely *Pyemotes* sp. on air dried weevil infesting cotton bolls. Washed moth, beetle or ant larvae or pupae are exposed to host seeking female mites, which rapidly colonized the insects and raised a generation in one week at 25±1°C and 60 per cent relative humidity (Weiser, 1963; Bruce, 1983). Bruce (1989) used larvae/pupae of cigarette beetle, *Lasioderma serricorne* to produced 223-239 off springs per female with in seven days of inoculation. *B. tarsalis* completed a generation in eight days when reared on live eggs of *Ephestia cautella* (Haines, 1981). It readily feeds on eggs and first instar larvae of *Tribolium* species, which can be used as an alternative host for its mass rearing. *A. lacunatus* is another important parasite associated with *Rhizopertha dominica* and other stored insects. Cultures are reared on *R. dominica* eggs on which seventeen off springs per female are produced (Faroni *et al.*, 2001).

Among the predatory mites, *Acaropsis docta* and *A. sollers* cultures are reared on *Trogoderma* eggs (Bhanot, 2003). *Cheyletus eruditus* cultures are reared on acarid mite, *Acarus siro* on which it completed its generation in four weeks (Zdarkova, 1986). *C. chinensis* is an another stored product predatory mite whose cultures is reared in the laboratory on *T. putrescentiae* (Kumud, 1987).

11.3 Alternate Diets for Mite Rearing

The evolution of artificial diets has had a major impact on the development of applied acarology. It relieves researchers from the trouble and expense of maintaining greenhouse facilities, host plants for rearing phytophagous mites and natural hosts of parasitic and predatory mites. To have a constant, economical production of vigorous competitive mites, a greater understanding about feeding is needed for each phase of the development cycle in relation to behaviour, preference, nutritional requirements etc. The diet should ensure proper nutrition at the lowest possible cost and yet be practical to use.

Development and use of artificial diets and ovipositional stimulants increased partly due to the emphasis on use of parasites and predators in biocontrol programme where augmentation of species and screening of pesticides is considered essential (Kantaratanakul and Rodriguez, 1979; House, 1980; Nottles *et al.*, 1982; Van der Geest *et al.*, 1983).

11.3.1 House Dust Mites

House dust mites are cultured on skin scales or human beard shavings but addition of yeast in diet fastened their development. Van Bronswijk (1972) reared the house dust mites on yeast in combination with human skin scales/wheat germ flakes. Taylor (1975) used yeast granules and human beard shavings with adherent skin scales defatted with acetone for mass culturing of these mites. Sodium caseinate is added to the diet proposed by Taylor (1975) for enhanced development and oviposition of house dust mites (Nannelli *et al.*, 1983). *Dermatophagoides farinae* and *D. pteronyssinus* are best cultured on dried yeast alone and beard shavings or insect meal was most suitable for *Euroglyphus maynei* (Hart and LeMerdy, 1988). Castagnoli *et al.* (1989) successfully mass reared these mites on a mixture of six parts of commercial rabbit food pellets, three parts of debittered brewer's yeast and one part of pork fat.

11.3.2 Plant Predatory Mites

Phytoseiids are the best-known predators among the Acari, which feeds on small arthropods (including phytophagous mites), pollen, honeydew, plant parasitic fungi and rarely on plant juices (Huffaker *et al.*, 1976; Overmeer, 1985; Sabelis, 1985; McMurtry and Rodriguez, 1987). Several reports indicated the successful rearing of phytoseiid on different pollen diets (Putman, 1962; Overmeer, 1981; Duso and Camporese, 1991; Bowie *et al.*, 1999).

Depending on the extent of pollen use or required by a species, different phytoseiids have been classified into four functional groups. Some phytoseiids require pollen for successful development and reproduction (*e.g. Euseius*), some use it as secondary part of a generalized diet (*T. pyri*), others rarely or never consume pollen (*P. persimilis*). Phytoseiid mites feed on pollen by picking up individual grains, cracking them open and sucking out the contents. Feeding rate of 100 grains per hour have been observed for starved mites. Since the nutritional value of pollen varies across plant species, accordingly developmental and reproductive responses of phytoseiids to different pollen also varies (Tanigoshi *et al.*, 1984). *T. pyri* readily feeds and reproduces on variety of pollen but tend to do better on pollen from trees and grasses that flower in the spring. Its reproductive rate can be as high on some types of pollen as on spider mite prey (Zaher *et al.*, 1975; 1984). Positive correlation is achieved between the peak in available pollen and population increase of phytoseiid mites (Grafton *et al.*, 1995; Addison *et al.*, 2000).

A. concordis is reared on castor pollen which is reported as most preferred food as compared to tetranychids, tenuipalpids, eriophyids (Nageschandra *et al.*, 1999). Castor pollen is also found as favourable diet for *A. ovalis* although generation is successfully completed on cotton pollen and on castor pollen mixed with egg (Borah and Rai, 1989). Manjunatha *et al.* (1995) observed castor pollen with honeydew and yeast as most favourable diet for its multiplication. Successful rearing of *A. teke* (employed for control of the cassava green mite, *Mononychellus tanajioa*) for twenty-five generations on an artificial diet is achieved by Ochieng *et al.*, (1987). Thirty one diets based on sugar or pollens are evaluated by James (1989) for their effect on development, survival and oviposition of *A. victoriensis*. Longevity of adults is enhanced on sugar granules or sugar based diets as majority live for more than four weeks and a few survived for 10 weeks. Oviposition (0.1 eggs/day) is found to be least on this diet as compared to natural prey or pollen diet. Pollen from rose and gum tree are recorded as most suitable diet, which stimulated the oviposition (1.5 eggs/day) along with 100 per cent survival of predatory population. Freshly collected pollens are made available to mites by providing two anthers per arena, which are replaced at 48-72h intervals. White and Laing (1977) used apple pollen for rearing of *Z. mali* (Stigmaeidae) from egg to adult. The species is also successfully reared on grape leaf tissue and yeast (Santos, 1991).

11.3.3 Storage Mites

Although it is convenient to rear acarid mites on wheat flour, adding one part of yeast in four parts of wheat flour increases the oviposition and causes rapid development. In six species, egg production is directly correlated with the type of food consumed (Woodring, 1969). In addition to oviposition, longevity of adults is also dependent on food. Females and males feeding on wheat germ survived for 13 and 16 days, however, longevity is recorded as 21 and 32 days, respectively, on yeast. Pankiewicz-Nowicka *et al.* (1982) explained the attractiveness of wheat germ as due to a mixture containing fats and biologically active components and not to the proteins, free amino acids and carbohydrates alone.

Pioneering work in introducing the artificial diet for acarid mite was done by Bot and Meyer (1969). They concluded that diet containing methyl p-hydroxy benzoate was best for *T. putrescentiae* and *R. echinopus*. Both the species survived better in diets containing mineral salts, carbohydrates, fats, vitamins and some amino acids (Boczek, 1975).

The above mentioned species and *G. domesticus* could utilize the lipid content of a number of substrates (ligament, tendon, feather and horn) whereas, they are unable to digest the proteinoids (collagen, casein, keratin). *A. siro* is able to digest the casein as synthetic diets based on casein and saccharose prolonged the life of female. Adults of both sexes survived longer on rye germs with supplements of vitamins and mineral salts (Boczek and Czajkowska, 1968; 1973). Studies on *L. konoi* revealed the increased number of hypopus formation in nutrient deficient diets.

The effectiveness of synthetic food is also tested on *Caloglyphus* species. Haydak diet (Haydak, 1936) developed for culturing wax moth larvae (*Galleria* sp.) and other insects, is found to be highly suitable for culturing *Tyrophagus* species. The diet consists of two components–Part I consists of corn flour: wheat flour: skim milk powder: dried yeast powder: wheat bran (4:2:2:1:2; w/w). Part-II is prepared by mixing equals parts of honey and glycerine. Equal parts of Part I and II are mixed thoroughly and kept for 24 hours for penetration of the liquid into dry component. Somchaudhary and Mukherjee (1971a; b) observed that Haydak diet infested with insects for more than a month is better than the freshly prepared one.

Salt concentration in food products also affects the development and interfered with culturing. Diets containing 18 to 21 per cent tricalcium phosphate reduced the *T. putrescenteae* fecundity by 50 per cent, whereas, 1.5 per cent concentration of salt had no effect on egg viability and post embryonic development (Ignatowicz and Pankiewicz-Nowicka, 1980). The latter concentration (1.5 per cent) of the salt, however, reduced the *A. siro* fecundity by 75 per cent with no effect on longevity (Ignatowicz, 1980). The development of acarids is not suppressed by increased mineral content in food upto 6 per cent.

Among the parasitic mites, semi-synthetic diet for *P. tritici* was developed by Bruce (1989). Yeast hydrolysate (2500 mg), casein hydrolysate (2500 mg), Wesson's salt mixture (400 mg), cholesterol (5 mg), sucrose (7000 mg), RNA solution (10 ml) and Tween 80 (0.1 ml of 10 per cent solution) are added to 70 ml distilled water and heated at 80°C for thirty minutes with regular stirring. Stock is cooled to 60°C and Vanderzant's vitamin mixture (400 mg), gelatin (500 mg; Gel strength 300 bloom) are added and stirred for ten minutes. Stock is again cooled to 40°C, streptomycin sulphate is added and stirred for five minutes. Stock is refrigerated until needed. Just prior to use, in 18.75 ml of stock solution, 6.25 ml chicken egg yolk is added and adjusted to a pH of 6.5-7.0 with potassium hydroxide. Each diet bag contained 1 ml of diet and sealed. Two bags per petri dish containing female mites are released and petri dish is ringed with petroleum jelly. Cultures are maintained at 26±1°C and 85 per cent relative humidity.

11.4 Quality Control in Mass Rearing

The science and art of mite rearing is entering into complexity and sophistication as the need has grown for mites research and its management. Genetic variability is important in rearing programmes of bioagents to ensure the production of competitive individuals that affect target population as planned. So mating competitiveness, pesticide resistance and the ability to seek out their hosts is must for predatory/ parasitic mites. In quality control, morphological, physiological (time of development, temperature adaptation, disease and pesticide resistance, sex ratio) and behavioural (phototaxis, dispersal, host preference, mating behaviour, egg laying capacity) traits are modified according to need and this could be achieved through genetic improvement. Genetic improvement through artificial selection of desired characteristics, hybridization and genetic engineering has great potential (Hoy, 1976; Beckendrof and Hoy, 1985) in Acarology.

A selection programme must begin with basic information about the organism, the traits to be selected and the knowledge of how the trait is inherited *i.e.* whether it is controlled by one gene or more than one gene. Introduction of desired gene(s) into a species is possible either by transferring the gene controlling the trait from one species to another or by modifying a known gene *in vitro* and returning into the species of origin. Pre introductory evaluation of natural enemies involved certain quality control criteria like seasonal and internal synchronization with host, no negative effect, adaptation to climate, host specificity, high kill rate and good searching efficiency for inoculative releases. For inundative releases, adaptation to climate, no negative effect, good cultural method and searching efficiency are considered essential (Van Lenteren and Woets, 1988).

The following are the desirable traits which could be exploited to increase the quality of acarine bioagents in terms of greater efficiency in predation.

11.4.1 Morphological Traits

Phytoseilus species is a specialized predator of *Tetranychus* spp. having highly dense, complicated and irregular web. Morphological adaptation by having comparatively large body size and long setae in the mid-dorsal and lateral part of the body shield have made them more efficient to move through the thick stranded webs of spider mites (Sabelis, 1981; Sabelis and Bakker, 1992). These morphological traits of specialized predators (*Phytoseilus*) of mite pests could be exploited in improving the generalist predators (*Amblyseius, Typhlodromus Metaseiulus*), which at present show unfavourable morphological traits to move in the dense web of spider mites. Generalist predators are preferred because of their omnivorous nature. They can survive and reproduce on pollen, leaf sap, small insects in addition to pest mites, thus regulating the pest population over a longer period of time. As they can be easily reared on pollen/ artificial diet (McMurtry and Seriven, 1966), it would be more economical to build predatory population by dusting the pollen as a supplementary food on the prey mite infested plant (Ghosh, 2001).

11.4.2 Physiological Traits

Pesticide resistance has been studied in greater detail than any other trait in predatory families like Anystidae (MacPhee and Sanford, 1954), Ascidae (Graham, 1970), Bdellidae (Wallace, 1954), Cheyletidae (Strong *et al.*, 1959; Zdarkova and Horak, 1987). Erythraeidae (Hagley and Simpson, 1983), Hemisarcoptidae (MacPhee and Sanford, 1954), Macrochelidae (Axtell, 1966), Phytoseiidae (Penman *et al.*, 1981; Marwick, 1986; Hassan *et al.*, 1988; Gerson and Cohen, 1989; Marwick *et al.*, 1990), Stigmaeidae (Nelson *et al.*, 1973; Parent, 1961; Weires and Smith, 1978) and Tydeidae (Mendel and Gerson, 1982; Knop and Hoy, 1983).

The discovery and utilization of pesticide resistance in some phytoseiids has led to numerous pest control programmes with resistant predators at their core (Hoy, 1985).

Amblyseius fallacis is an important predator of spider mites (Thistlewood, 1991), which was found very susceptible to pyrethroids like other phytoseiids. Consequently, phytoseiids have been the subject of genetic improvement for resistance to pyrethroids using several methods involving field and laboratory selection (Hoy, 1985; Croft, 1990). Field selection increased the LC_{50} of *Typhlodromus pyri* (Fitzgerald and Solomon, 1992) and *A. fallacis* (Thistlewood *et al.*, 1992), against pyrethroids. Likewise, phytoseiids like *Metaseiulus occidentalis* have developed resistance against organophosphates, carbamates and sulphur compounds. Since, field selection is not always assured to augment naturally acquired resistance, artificial selection and crosses has been done on phytoseiids reared continuously in green houses and sprayed directly (Strickler and Croft, 1982; Petrushov, 1991) or introduced into a treated diet in laboratory (Hoy, 1983; Petrushov, 1992). Small population of selected stages of phytoseiids are selected using different bioassay techniques which have been described in separate chapter 'Bioassay techniques'. Selected individuals are reared normally on tetranychids. In each generation, known number of females are collected and challenged with insecticide to get resistant individuals which are again reared on normal food.

Carbaryl-OP strains, permethrin-OP strains of *M. occidentatis* have been developed which met the criteria for field success (Hoy, 1983). With time, these strains have not only spread, distributed, multiplied and survived in the apple, pear and almond orchards but made impact on the pest (spider mite) population in several parts of the world. Later, Beckendorf and Hoy (1985) developed carbaryl-OP-permethrin and carbaryl-OP-sulphur resistant strains through crosses and selection, which performed well in field.

Temperature adaptation is another physiological trait, which has attracted the attention of scientists. Storing insects or mites under low temperature to extend shelf life has been commonly practiced in integrated pest management programme (Archer *et al.*, 1973; Morewood, 1992). Cold storage is considered a valuable aspect of mass production. *A. fallacis* eggs, stored at 2-4°C for four days have no adverse effect on hatchability (Thistlewood *et al.*, 1992). Provision of moisture and food during cold storage greatly increased the survival of mobile stages of mites (Hamamura *et al.*, 1978). When food is provided, *P. persimilis* survival at 7.5°C is 97 per cent after four weeks and subsequent longevity and fecundity of these mites are comparable to mites taken directly from mass rearing cultures (Morewood, 1992). During cold storage, individual mites are held in petri dishes with moist filter paper and *T. urticae* eggs.

Deutonymphs and tritonymphs of *Halotydeus destructor* stored at 1.5°C in sealed plastic boxes remained more than 50 per cent after 12 days of storage. Adding fresh sub clover leaves into storage box enhanced their survival.

11.4.3 Behavioural Traits

Searching capability, dispersal powers and reproductive potential are the major behavioural traits of any successful acarine biocontrol agent (McMurtry, 1982; Rosen and Huffaker, 1983; Gerson and Smiley, 1990). The dispersal mechanism may be autonomous or insect mediated. Many predatory mites are fast runners, rapidly combining the immediate environment for food; some locate their prey by kairomones (Sabelis and Dicke, 1985). Bdellids appear to advance by walking (Currie, 1934) and manifest an ambushing strategy to tether their prey (Sorensen *et al.*, 1983). *Blattisocius tarsalis* utilizes stored product insects for dispersal and then feeds on their eggs (Haines, 1981). *Macrocheles* spp. employs the same strategy, using filth flies as vectors and food (Axtell and Rutz, 1986). *Cheyletus* spp. are introduced to stored product habitat by birds and small scavengers (Woodroffe, 1953). Lady beetles transport deutomymphal stages of *Hemisarcoptes* spp.

Some predaceous mites have surmounted the problem of reaching their discrete prey habitats by being phoretic on insects, which may themselves serve as food (*B. tarsalis, Macrocheles* spp.) or feed on the same pest (*Hemisarcoptes* spp.). These traits could be exploited for successful biological control programme. When food is scarce and wind currents are present, *Neoseiulus fallacis* assumed a raised stand on leaf margins (Johnson and Croft, 1976). This leads to down wind dispersal for a distance that depends on the elevation from which they arise (Johnson and Croft, 1979). In addition to wind direction, temperature, density of spider mites, plant density affects the dispersal.

Coop and Croft (1995) studied the *N. fallacis* dispersal behaviour in strawberry field. They reported that 100 adult females for 1-2 m of row can be released after *T. urticae* have achieved 2-5 female adults per leaf. They colonise upto 100 sq. m. around each release point and covered the entire field in a single season. For each point of release, they placed 15-30 bean plants carrying 100±20 female adults on to strawberry foliage.

The power of increase (reproductive potential) for majority of predators/parasitic mites is usually similar to or greater than that of the pests. Logan (1982) observed that the intrinsic rate of natural increase, of phytoseiids is often similar to that of their spider mite prey. *Cheyletus eruditus* develops somewhat faster than its acarid diet under slightly warmer and drier conditions, increasing the chances for successful pest control (Solomon, 1969). Species of *Blattisocius, Macrocheles* and *Hemisarcoptes* reproduce more rapidly than their prey (Haines, 1981; Gerson and Schneider, 1981; Axtell and Rutz. 1986).

11.5 Control of Contamination in Mite Rearing

Mass rearing of spider mites is frequently impeded due to contamination by various predators (phytoseiids). The rapid turn over of plants (Scriven and McMurtry, 1971), can rid the colony of most of the predators. Chilling the colony killed both host plant and spider mites before eradicating the unwanted phytoseiids but hot water treatment succeeded in killing phytoseiid mites (Ho and Chen, 1992). Immersion for 60 seconds in water of 50°C killed all stage of phytoseiid mites. Initial reduction in pest population is also noticed but reportedly they recovered soon and no adverse effect is reported on seedlings. To prevent recolonizatin of phytoseiids after the treatment, seedlings are advised to place in glass bottles half filled with water and bottles in water pan.

In stored mite (acarid) culture, high population density per unit area causes high moisture, which leads to fungal growth. Fungi are reported to reduce the fecundity of *Caloglyphus* and *Tyrophagus* due to production of mycotoxins (Rodrigquez *et al.*, 1984). Parkinson *et al.* (1991) reported that fungi shortened the *Acarus siro* male lives in culture, thereby reducing the strength of cultures. High humidity is also a problem in some plant and soil mite culture, which promotes excessive fungal growth. The fungus, *Hirsutella thompsonii* caused rust mite mortality (Reed *et al.*, 1964) and *Verticillium* sp. reduced the earth mite population in cultures. In latter case, increasing the air circulation with in the pots and between the pots can control the contamination. Due to low humidity, contamination through fungal spores is also reported in house dust mites where low relative humidity was the cause.

11.6 Release and Shipment

In the first ever international shipment of an acarine natural enemy (Howard, 1930), *Tyroglyphus phylloxerae* was shiped to Europe by Riley in 1874. Then, in 1902, *Pyemotes* species was introduced from Mexico to USA (Hunter and Hinds, 1904), but no data are available about the method(s) used in both the shipments.

Within Australia, from Western territories to eastern territories, Currie (1934) developed a method to send Bdellids. An open glass tube was plugged at both ends with sphagnum moss to provide required humidity. Inside the tubes, dry bark strips, clover leaves and collembolans as prey were released with Bdellids, which reached to their destination in good condition. Wallace and Walters (1974) replaced the sphagnum moss with saturated facial tissues for sending Bdellids from Australia to South Africa and got good results.

Hypopodal stage of *Hemisarcoptes* species was shipped to Bermudab by Bedford (1949) under the elytra of *Chilocorus* spp. Predator augmentation of *Cheyletus eruditus* @ 1:100, 1:1000 *Cheyletus: Acarus* was suggested by Pulpan and Verner (1965) and Zdarkova and Pulpan (1973). The former sieved out *C. eruditus* from predator rich substrates and released them on grains where predators were rare. The latter authors showed that predatory mite could be kept alive for several months in cold storage and then used successfully.

Among the soil mites, surface sterilized eggs of *Macrocheles peregrinus* were shipped to Australia from South Africa (Wallace and Holm, 1983) to control flies. Small scale releases of *Allothrombium pulvinum* to control aphids indicated that mite: aphid (3:1) could control these pests (Zhang, 1988).

Phytoseiid handling and shipment procedures are discussed by Overmeer (1985) and Herren *et al.* (1987). Before releasing predators, Herren *et al.* (1987) marked them with flourescent powders, which persisted for about twenty four hours. This technique enabled them to follow the spread of the mites. In India, *A. rhodendronicus* provided efficient control of *T. ludeni* in field conditions with in four weeks time (Krishnamoorthy, 1983).

Morewood (1992) discussed shipment procedure of *P. persimilis* release. Individual female mites in petri dishes with moist filter paper and *T. urticae* eggs were kept under cold storage. Prior to shipment, these petri dishes were loosely filled with wheat bran, which hold moisture well and provisioned with more *T. urticae* eggs on filter paper.

Release of one *A. tetranychivorous* per six *T. ludeni* checked the pest population with in ten days in okra (Nangia and Channabasavanna, 1983; 1995). Release of twenty *A. ovalis* mites per plant was effective in checking the thrips and mite population on chilli when initial count of pest population was 0.8 and 0.61 per three leaves, respectively (Manjunatha *et al.*, 1995).

11.7 Concluding Remarks

Cost of dietary ingredients and quality of particular strain is increasing in its importance in mite rearing programme. Diets that are more balanced in proteins, carbohydrates, lipids, vitamins and minerals are needed for more efficient utilization. To develop an alternate diet for any species, clear understanding of its behaviour and constituents of natural diet is must. *Pyemotes* sp. is an important biocontrol agent against stored insect pests. It probes and pierces the cuticle of host for feeding. To develop a suitable diet, an appropriate membrane is required to contain the diet and allow the feeding process to occur. Bruce (1983) achieved a breakthrough when he found a plastic wrap of appropriate thickness, strength and texture to be used as membrane. Similar studies are required for formulating new diets for other potential biocontrol agents (predatory and parasite) that cannot be reared at the present time.

For example, *Eutrombidium* (Welbourn, 1983) and *Dinothrombium* (Newell, 1979) which have high fecundity rate (47,000-80,000 eggs/female) are potential biocontrol agents of grasshoppers and termites. Larval stage is a parasite whereas nymphs and adults are predators on grasshoppers/termites. For detailed studies on their predatory efficiency, mass rearing techniques are pre-requisite which at present are not known.

Cannibalism in some phytoseiids (*e.g. Amblyseius, Phytoseilus*) require individual rearing or rearing at high prey densities which increased the cost of rearing. Improvement in existing diets and methods for acarine natural enemies (both predators and parasites) are required. Nutritional and ovipositional stimulant diet will yield more progenies per female. At present, a semi-synthetic diet for *P. tritici*, a parasite of stored insect pests provides 137-151 offsprings per female (Bruce, 1989) whereas a female produced 223-239 offsprings on natural host.

Quality testing is of paramount importance for mite species used in alternate methods of control. Mating competitiveness, rate of development, per cent survival and adult emergence, number of eggs produced and synchronization with pest population are some of the indicators which should be kept in mind to develop particular strain. Pheromone production, locomotor's activity, enzyme levels are some of the other variables which affect the performance. In addition, breeding programmes to develop strains of mites for special purposes like resistant strain against acaricide /insecticide, improved searching behaviour, seasonal adaptation, altered host preference and increased female progeny in population are the future needs in the changing agricultural scenario. Few pesticide resistant strains of phytoseiids are available but still there is a long way to go. Secondly, mass-produced acarine natural enemy populations have to be stored for varying periods so as to make them available for pest outbreaks. Cold storage is a valuable aspect of mass rearing which needs to be explored (Leopold, 1998; Lin and Ridsdill-Smith, 2000). Morewood (1992) refined the mass rearing systems of *P. persimilis* by successfully cold storing the mites with adequate food supply at 7.5°C for four to six weeks and with no effect on subsequent longevity and fecundity of mites. Such studies are needed for other acarine bioagents. Zdarkova and Pulpan (1973) successfully cold stored the *C. eruditus* for several weeks before releasing them at pest outbreak sites.

Developing quality tests for acarines will be another exciting area of future research. It was observed that laboratory reared individuals sometimes behave in unexpected ways when released in field. This may be due to the differences in rearing conditions and hosts. Rearing at normal environmental conditions and exploration of native acarine bioagent will probably solve this problem by some degree. A day will come when success in mite rearing will be measured not only in empirical terms but both in terms of quantity as well as quality per unit of cost.

Chapter 12

Counting

The variations in mite size, habits and habitats as well as the objectives for which the evaluation appraised is useful, determines the method to estimate population density. The small size, high mobility and large numbers on host make accurate counting difficult, thus a number of different methods have been developed for estimating mite densities in various habitats. Many a times, it becomes impracticable to count all the mites present in the sample depending on the nature of sample and level of infestation. To overcome such difficulties, mostly, numbers of mites are counted from known quantity of four to five sub-samples from each sample.

12.1 House Dust Mites

Counting methods described under soil and stored mites counting are used for house dust mite counting also. Since sampling is usually done through collection of dust, excess dust can be removed through sieving/ floatation before actual counting is done.

12.2 Parasite Mites

12.2.1 Counting Dish Method

Square counting dish is used to count parasitic mites from hive debris or mites collected from animals through fumigation technique. Average numbers of mites are calculated by counting the mites in each square (Figure 12.1).

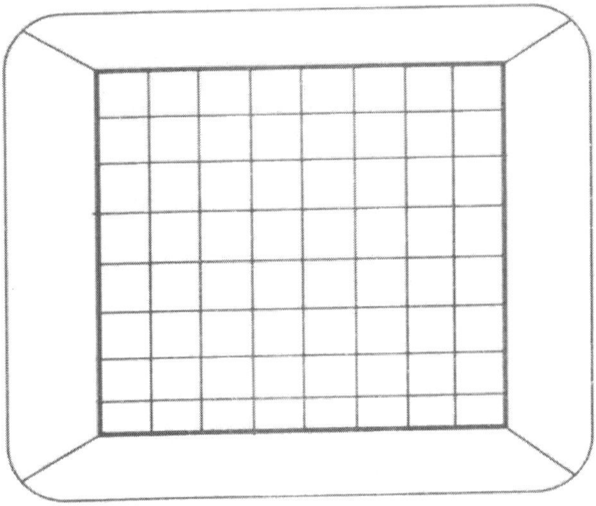

Figure 12.1: Counting Dish

12.2.2 PTMIS Method

McClure (1989) described the method to compare the intensity of infestation on individual feathers, which was later modified by Behnke *et al.* (1995). This is a rapid method of counting without disturbing or distressing the birds. Left wing of the bird is extended and held up to ambient light. The complete length of the primary feather is exposed and primary covets and under wing covets are moved gently aside to examine the mite population.

On the basis of mite infestation intensity, a subjective scoring system is developed in the range of 0-3.

0: Represents feathers which are not detectably infested;

1: Represents feathers with some mites congregating around the feather shaft (usually mid shaft/base) and little or no spread to barbs;

2: Represents feathers in which mites are observed packing the feather shaft and extending along the barbs;

3: Represents feathers in which mites are observed packing the feather shaft and the barbs.

Each feather is given a score and total is counted by adding all individual scores, which is termed as primary feather total mite infestation score (PTMIS). Actual mite counting done under microscope on one primary feather sampled per bird should calibrate the score. This method has an advantage over other methods of counting as it determines the total mite infestation on live birds without rendering harm to the birds themselves.

12.3 Plant Mites

As most of the methods involved removal of plant part, few were of the opinion that at low densities, it affected the dynamics of interest so direct counting should be done with the plant parts attached to the plants. Few scientists have tried to move a microscope from leaf to leaf in the laboratory (Takafuji and Chant, 1976) or in the field (Lawson and Walde, 1993) or use of hand lens (Walde, 1991).

12.3.1 Direct Counting

This is one of the simplest methods, which involves the removal of plant part and to count the mites under a binocular microscope. This method is considered to be most precise in counting plant mites. Mite infested leaves and twigs are collected from the crop/field and actual counts of all stages present are made on these samples under binocular microscope. It is also a time consuming procedure. Accuracy depends on the time interval between removal of plant part and examination and the degree of errors that occurred due to mites' movement. Taking subsamples of mite infested leaves can reduce the error. Also, samples are sometime cooled or sprayed with a fixative immediately after sampling to reduce this problem.

12.3.2 Imprint Method

Mite infested leaf is placed between two papers of proper absorptiveness and mites are crushed against the papers. Imprints of all stages of mites are recorded in this method followed by counting. Accuracy of this method depends on the distinctive colours of different species. Different species can, with practice, be distinguished but assessment of age structure is usually not possible. Secondly, if two species of same colour are present on the leaves, differentiation becomes impossible after crushing. At high densities, the imprints can coalesce, making the count inaccurate.

12.3.3 Field Counting

With the help of magnifying glass, rapid counting of adult mites is possible without actually disturbing them from their natural breeding sites on the host plant. In case of heavy infestation, counts by tens are feasible.

12.3.4 Mite Brushing Machine

This is one of the popular options in mite counting and was described by Henderson and McBurnie (1943). Leaves are passed through a set of brushes and the dislodged mites fall on to a rotating glass disk with a sticky coating, which is then examined under a binocular microscope. Species and stages are better identified but all of the mites are not dislodged from the leaf.

12.3.5 Plant Part Washing

This is also considered a reliable method in which plant parts are washed with hypochlorite solution to wash off the mites, which are collected in a container. The mites are then separated on filter paper and counting is done under microscope.

12.3.6 Per Square cm Counting

This is an improvised method of direct counting which is used popularly to count phytophagous mites except eriophyid mites. On a mite infested leaf, per square cm area are marked on 3-4 places, number of mites and its stages are counted under binocular microscope using tally counter. In case of heavy infestation, counting should be carried out on both dorsal and ventral surface.

Three leaves are plucked from top, middle and bottom portion of each plant in a labelled polythene bag for counting under microscope in laboratory.

12.3.7 Per Leaf Counting

Phytoseiids and other predatory mites are counted with this method. These fast moving mites are generally encountered near the mid rib, veins of leaves and direct counting per leaf is made under the microscope.

12.3.8 Bud Mite Counting

Buds are macerated before centrifuging the bud tissues. Separated mites from tissues are then counted (Sternlicht, 1966).

12.3.9 Eriophyid Counting

This method was described by Janardhan *et al.* (1972) on pigeonpea. Mite infested leaves are immersed in methanol/ethanol

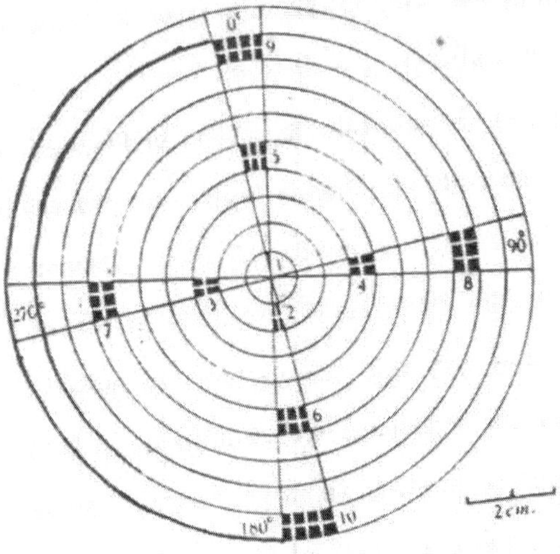

Figure 12.2: Card Disc Method (Solomon, 1972)

or acetone + glycerol (10:1) for three to four hours. Leaves are then gently spread on a glass slide to count the number of mites.

12.3.10 Counting by Photographs

Mite counting is also done after photographic sampling of leaves as discussed in Chapter 'Collections and Extractions'. Different mite species and stages are marked with different colours or symbols to make the counting easier.

As most of the methods involve removal of plant part, few scientists are of the opinion that at low densities, it affects the dynamics of interest so direct counting should be done with the plant parts attached to the plants. Few scientists have tried to move a microscope from leaf to leaf in the laboratory (Takafuji and Chant, 1976) or in the field (Lawson and Walde, 1993) or have used hand lens (Walde, 1991).

12.4 Soil and Storage Mites

12.4.1 Counting through Counting Dish

Plastic counting dish with horizontal and vertical lines are used in stored, soil and parasitic mite samples. Mites extracted in water through floatation technique or Berlese funnel method or with other technique are transferred to counting dish (Figure 12.1), which is placed under the microscope. Few seconds are given to mites to settle and then one square of the counting dish from one side is adjusted under microscope. Numbers of mites are recorded from that particular square using tally counter before proceeding to next square occupied by mites.

In absence of counting dish, draw one mm squares on a plain paper and paste it on the bottom of petri dish or petri plate, count the number of mites in each square with the help of tally counter.

In case of dry sample, known quantity of sample is uniformly distributed in the counting dish before counting the mites. Mites and its developmental stages should be counted from all the squares occupied by them.

11.4.2 Card Disc Method

Solomon (1972) gave this method to count number of mites in stored grains. A card disc (12-15 cm) is prepared by making four sectors (12°) from the centre in position of 0, 90, 180 and 270° (Figure 12.2). A number of concentric circles of diameter x, 2x, 3x (x=0.6 cm) etc. are drawn which cut the each section into a series of segments. Then, proceeding from the centre outwards, the segments are marked from the sectors at 0, 90, 180 and 270°. The disc is placed under the petridish containing mite-infested sample and mites are counted in marked segments. The total is then multiplied by 30 that give the number of mites in the dish.

Chapter 13

Bioassay

Bioassay, in case of chemicals, refers to a method in which property of a chemical is measured in terms of some biological response. It is used for the detection of minute quantities of insecticides on a quantitative basis. It is a simple, versatile, easy and sensitive technique for determining residues of a wide range of toxicants. In case of acarines, bioassay techniques are employed to determine susceptibility and resistant levels to acaricides.

The development of resistance depends on the interaction of many factors *e.g.* the degree of dominance and fitness of the genes confirming resistance, the breeding habits of the pest, extent of exposure to the toxicant and other ecological factors. Four main mechanisms of resistance are recognized:

1. Behavioural changes which reduces the likelihood of the pest becoming exposed to the pesticide.
2. Reduced penetration of the pesticide by thickening of the cuticle.
3. Enhanced breakdown of the pesticide due to the possession of highly active or concentrated detoxification enzymes.
4. Modifications of the pesticide action sites that leads to a reduced sensitivity of the susceptible systems of the targets.

The purpose of bioassay studies is to evaluate the toxicity of selected acaricide/insecticide with diverse mode of action to the same species. In an effective bioassay, the indicator species should be sufficiently sensitive to detect even small amounts of acaricides and should express the response with increasing acaricidal concentrations. In studies to determine the innate susceptibility of living organism to pesticide, the most sensitive bioassay is frequently the technique of choice. The 48h continuous exposure was found to be more sensitive than five second dip technique in terms of lower LC_{50} values for comparable acaricides.

Like in entomology, several methods are employed in acarology for bioassay studies. Various designs of mite cages suitable for bioassay work have been described which varies with the test organism. Mites of known age and of both sexes should be used in bioassay studies.

13.1 Topical Application

In this method, groups of known number of mites are placed under potter's spray tower directly, from which spray is done at constant pressure for fixed time interval. Mortality count is done after 24/48 hours. Reliability of this technique for fast moving mites and mites under copious webbing is relatively poor.

13.2 Spray Method

Mites are sprayed on absorbent paper and both the mites and paper are transferred to petri dish (Chen and Lo, 1989). Mortality is counted after 48h. Alternatively, in case of plant mites, leaf discs are sprayed with insecticide using a potter spray tower that produced a known weight of deposit after a fixed settling time. The sprayed discs are maintained under optimum conditions on moistened cotton wool for 48h (Edge and James, 1982; Markwick *et al.*, 1990).

13.3 Petri Dish Method (Dry Film Technique)

Organism is exposed to a film of the toxicant in this technique. Thistlewood *et al.* (1992) discussed this technique for plant mites, but this could be applied for other groups of mites also. One ml of acaricide/insecticide solution is placed on the bottom of petri dishes, the dishes are closed and the solution swirled briefly in upright and inverted positions. The dishes are opened, excess solution poured off and the deposits are allowed to air dry for sixty minutes. Control dishes are treated with solvent in the some way. If not in use, dishes could be stored at 2°C and used with in twelve days. Open dishes are placed briefly on chilled freezer packs. Known numbers of mites are released in each dish before closing them tightly and mortality is counted after twenty-four hours.

13.4 Leaf Dip Method

This technique is used for detecting the ovicidal action of acaricide. Leaf discs of 15 mm diameter are cut from fresh leaves and are placed on wet filter paper with lower surface up, on a saturated cotton wool in a shallow tray. Known numbers of gravid females or pairs are released on each disc for twenty-four hours at 28°C and 50 per cent relative humidity. Then, females/pairs are removed from the disc using a fine brush and discs are transferred to acrylic plates covered with filter paper suspended in water. Number of eggs on each disc is calculated.

Serial dilutions of acaricide are prepared in water in which 0.03 per cent wetting agent is added. Discs are dipped in the solution for five seconds with gentle agitation. Excess liquid is removed by touching the edge of the disc on to absorbent paper before returning the disc to the plate. Treated discs are held at 28°C and 50 per cent relative humidity for seven days. Unhatched egg and dead larvae are counted for calculating per cent mortality in each disc. The discriminating dose is calculated from dose response line after probit analysis.

13.5 Slide Dip Method

Fresh dilutions of acaricide in water: acetone (1:1 v/v) are prepared from stock solutions in acetone. A piece of double sided adhesive tape is affixed to microscope slides and a second piece of adhesive tape is placed on the other side, sticking side up (Anonymous, 1984; Thisthewood *et al.*, 1992).

Adults are placed on their backs on each slide using a fine sable brush. Slides are dipped so that mites are immersed completely in solution and gently agitated the slide for five seconds to ensure complete wetting. Mites similarly dipped in 1:1 water: acetone solution acts as control. When withdrawn, slides are touched down on their edge on an absorbent paper towel and allowed to dry at room temperature for fifteen minutes, then transfer to controlled environment chamber. After twenty-four hours, mortality is counted.

Knowles *et al.* (1988) also used this technique to examine the toxicity of 64-acaricide/ insecticides against *R. echinopus* (stored mite) and spider mite.

13.6 Single Leaf Disc Technique

The mites are exposed to acaricide residue on a small treated leaf disc cut from host substrate. Sprayed leaf discs are allowed to dry for six hours before use. A barrier of moist cotton is placed around the circumference of leaf disc, which prevent mites from escaping. This technique was also used by Roush and Hoy (1978) and Hoy and Knop (1981) for phytoseiid mites. For phytoseiids, Avella *et al.* (1985) used haemolysis tubes for bioassay studies.

However, this technique has some limitations. In some cases, it is difficult to confine mites like *P. ulmi* and *T. occidentalis* who have a tendency to wander into the barrier, where they perish. Moreover, sometimes due to the repellent nature of acaricide, mites leave the treated area and escape into the surrounding water.

13.7 Closed Double Leaf Disc Technique

This technique is a modified version of single leaf disc technique, developed by Riedl and Shearer (1991). Mites are caged between two leaf discs, which are held together by spring-loaded clips. The larger leaf disc (23 mm diameter) is positioned over a strip of filter paper, then the smaller leaf disc (20 mm diameter) and the plastic ring is centered over it. Filter paper, leaf discs (with sprayed sides facing each other) and plastic ring are positioned in the clip unit (Figure 13.1). The clamping action of spring-loaded clips pressed the plastic ring against the metal washer, which provides a tight seal around the edges of the two leaf discs.

13.8 Impregnated Plastic Strips Bioassay

This bioassay technique is generally applied to honeybee mites or stored grain mites. Impregnated plastic strips or acaricide / insecticide applied on plywood boards are placed with in the hive/ storage container for longer duration (six weeks).

Since, in this method mite is exposed continuously to the toxicant, it is useful in selecting resistant individuals for further research (Mozes-Koch *et al.*, 2000).

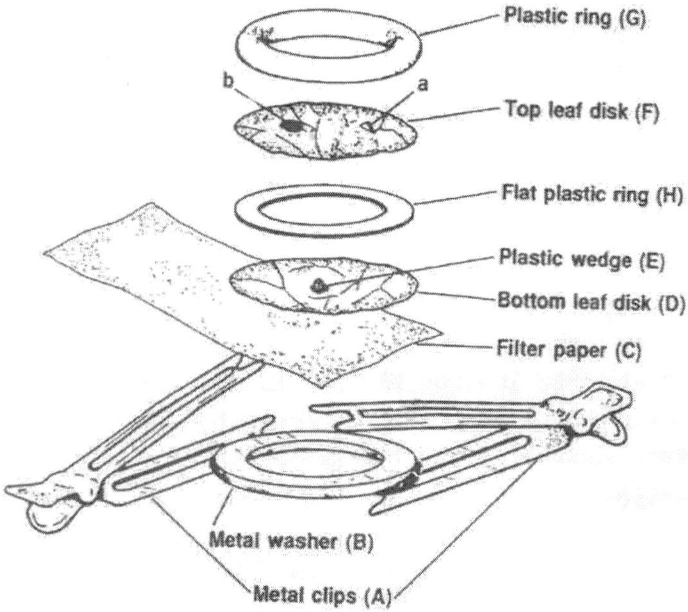

Figure 13.1: Closed Double Leaf Disc Technique (Riedl and Shearer, 1991)

13.9 Impregnated Petri Dish Bioassay

Petri dishes are coated with paraffin into which pesticide is added (Milani, 1995; Mozes-Koch *et al.*, 2000). Group of known number of mites are introduced on to the coated plates, which are placed in incubator at optimum condition for six hours and allowed the mites to move freely inside the petri dishes. The survivors are then, transferred to paraffin coated petri dishes without insecticides for twenty-four hours. The mites are offered their choicest food. Dead mites are counted under microscope for calculating percent mortality.

Chen and Lo (1989) used the similar type of technique in which mites are placed on the surface of an artificial medium that had been treated with the test compound and termed the technique as 'integrate method'.

13.10 Impregnated Paper Bioassay

Mites are exposed to pesticide in special paper chamber (Thind and Muggleton, 1998; Szlendak *et al.*, 2000). The chamber is made of two pieces of Whatman No. 1 (42.5 mm diameter) filter paper. One filter paper is moulded to create a 2 mm deep and 33 mm diameter depression. Second paper is unmoulded and used as chamber lid. Papers are secured by using glue between two squares of glass (75×75×4 mm), in the centre of which a 35 mm diameter hole is drilled (Figure 13.2).

Each filter paper is treated with known dose of acaricide solution. Control contains only solvent solution. Known numbr of mites are used in this bioassay. This technique is useful for various groups of acarines.

13.11 Pipette Technique

This technique was devised for conducting bioassay studies on parasitic mites (Foulk and Matthysse, 1964; Matthysse *et al.*, 1975; Hall *et al.*, 1978; Arthur and Axtell, 1983; Fletcher and Axtell, 1991). Disposable Pasteur pipettes (0.5 cm inside diameter × 11 cm length) are used in which fine mesh cloth is attached to the large end by a latex band. These pipettes are immersed down in acaricide solution

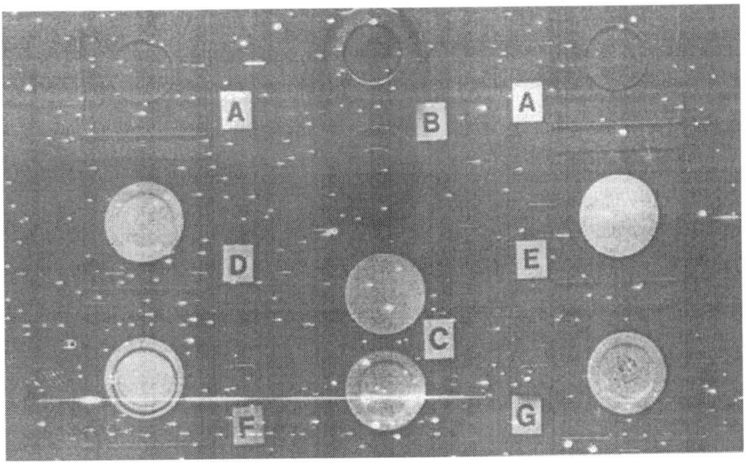

Figure 13.2: Impregnated Paper Bioassay (Thind and Muggleton, 1998)

from the cloth end side for sixty seconds. Control pipettes are immersed in solvent only.

After drying the pipettes for twelve hours, known numbers of adult mites are aspirated from collection container into a pipette and the narrow end is plugged with modeling clay. Pipettes are held in glass desiccators jars under constant indirect light at 27°C and 80 per cent relative humidity (super saturated solution of Na_2HPO_4). Mortality is determined after 24 and 48h, respectively. Each pipette is monitored under microscope and mites are considered dead if they are unable to move or had only a slight twitching motion of legs.

Zeman and Zelegny (1985) used a similar treated tube technique but removed the mites after twenty-four hours and held them in clean tubes for additional twenty-four hours before determining mortality, however, the results were comparable with original technique. Crystal and DeMilo (1984) tested acaricide residues on filter paper against parasitic mites.

13.12 Filter Paper Bioassay

This technique was described by Kuwahara *et al.* (1985) and Kuwahara (1986) for bioassay studies against stored mite, *R. robini.* Approximately 0.25 ml of acetone containing the desired concentration of acaricide is added to a glass scintillation vial (20 ml) containing a glass fibre filter (Whatman) and the acetone is allowed to evaporate. After adding 0.4 ml of distilled water, adult mites are transferred to the vial and vial is placed in the dark at 26±1°C. Control is prepared in the same way but without acaricide. Mortality is counted after 48 hours. In a continuous exposure technique like this, mites accumulated higher internal level of toxicant by both contact and ingestion, which could lead to harvesting of resistant individuals of a species at a later stage.

Chen (1990), Zhao *et al.* (1996) and Gencsoylu *et al.* (1998) also used this technique to estimate acaricides toxicity against various acarines.

13.13 Oral Feeding Technique

A polythene tube (0.017 mm thick) is filled with chemical solution (Yamada *et al.*, 1986). The surface of the tube is surrounded with tangle food. Known numbers of adults are inoculated and allowed to suck the toxicant through the tube. Tube is placed in desiccator of desired humidity and under optimum conditions.

DOSE RESPONSE CURVE

The mortality data generated against certain doses with response varying from 10-90 per cent is subjected to certain calculations in order to determine the unit of response. In order to draw reliable results, the per cent mortality obtained by bioassay technique is converted into probality mortality and doses. A straight line is achieved between both the factors.

A number of methods are available for computing the LD_{50} from the simplest method of Wrongstedt-Behrens to the most tedious but precise method of probit analysis of Finney (1971). Both the methods are given here for the benefit of readers.

1. Wragstedt-Behrens Method

This is the simplest method used to interpret the scanty data.

$$\text{Log } LD_{50} = \text{Log A} + \frac{50 - a \times \log 2}{b - a}$$

where,

A: Dose of insecticide which has a per cent kill immediately below 50 per cent mortality.

a: Per cent kill observed immediately below 50 per cent mortality

b: Per cent kill observed immediately below 50 per cent mortality

Log 2: Constant (0.301).

This method works best when the test organisms are employed in small numbers over a larger dose of fairly close doses. Only disadvantages is that it may tend to under estimation of LD_{50} if in between data is missing.

2. Probit Analysis

The following procedure using Maximum Likelihood Method (Finney, 1971) should be adopted for probit analysis using Table 13.1. Most of the columns are self-explanatory. Net percentage kill is calculated by Abbot's formula

$$\text{Net percentage kill (p)} = \frac{\%\text{mortality observed} - \%\text{ mortality in control}}{100 - \%\text{mortality in control}} \times 100$$

Empirical probit is recorded from Table I of Finney. Then, a graph between empirical probit and log of dose (x) is plotted to record the provisional probit (y) from a straight provisional line. For each provisional probit, weighting coefficient (w) is recorded from Table II of Finney and enter it in column 9. Corresponding to each p working probit (y) is also recorded from Table IV of Finney. The subsequent calculations are then carried out in the data sheet (Table 13.1) and should be taken to three decimal places.

Check

1=

Σnw

$Y = \Sigma nwy/\Sigma nw$; $x = \Sigma nwx/\Sigma nw$; $\Sigma x^2 = \Sigma nwx^2 - (\Sigma nwx)^2/\Sigma nw$

$\Sigma xy = \Sigma nwxy - (\Sigma nwx)(\Sigma nwy)/\Sigma nw$

$\Sigma y^2 = \Sigma nwy^2 - (\Sigma nwy)^2/\Sigma nw$

$x^2 = \Sigma y^2 - (\Sigma xy)^2/\Sigma x^2$

If homogenous, t = normal deviate for desired probality level (Pat 0.05 = 1.96), K-2 degree of freedoms.

Theoretical x2 =————

If heterogeneous, multiply all variances indicated by brackets and * (*) by heterogeneity factor (x2/k-2) and all 't' values must be for k-2 degrees of freedom.

(dose mortality curve) $b = \Sigma xy/\Sigma x^2 = \ldots\ldots\ldots\ldots\ldots \pm \sqrt{\Sigma}\,(1)^*/\,\Sigma x^2$

$y = y + b\,(x - x) = \ldots\ldots \pm \sqrt{\Sigma}\,1/\Sigma nw + (x\text{-}x)^*/\,\Sigma x^2$

at x (), y = ... ± ...... with range from ...... to ......

at x (), y = ... ± ...... with range from ...... to ......

at x (), y = ... ± ...... with range from ...... to ......

---

at x (), y = ... ± ...... with range from ...... to ......

at x (), y = ... ± ...... with range from ...... to ......

$M = x = (Y\text{-}y)/b = \ldots\ldots \pm \sqrt{\Sigma}1/b^2\,(1/?nw + (m\text{-}x)/\sqrt{\Sigma x^2})^*$

$m = (\text{Log } LD_{50}) = \ldots\ldots \pm \ldots\ldots$ with range from ...... to ......

Table 13.1: Data Sheet for Probit Analysis Using Maximum Likelihood Method (Finney, 1971)

Dose or Conc.	*Log of Dose (x)*	*Total Mites (n)*	*Observed kill (x)*	*% kill (p)*	*Net % kill (p)*	*Emp. Probit*	*Prov. Probit (y)*	*Wting coefficient (w)*	*Working probit (y)*	*(n) (w) = nw*	*(nw) (x) = nwx*	*(nw) (y) = nwy*	*(nwx) = nwx²*	*(nwx) (y) = nwxy*	*(nwy) (y) = nwyz*
—	—	—	—	—	—	—	—	—	—	—	—	—	—	—	—
—	—	—	—	—	—	—	—	—	—	—	—	—	—	—	—
—	—	—	—	—	—	—	—	—	—	—	—	—	—	—	—
—	—	—	—	—	—	—	—	—	—	—	—	—	—	—	—
—	—	—	—	—	—	—	—	—	—	—	—	—	—	—	—
—	—	—	—	—	—	—	—	—	—	—	—	—	—	—	—
—	—	—	—	—	—	—	—	—	—	—	—	—	—	—	—
—	—	—	—	—	—	—	—	—	—	—	—	—	—	—	—
—	—	—	—	—	—	—	—	—	—	—	—	—	—	—	—
—	—	—	—	—	—	—	—	—	—	—	—	—	—	—	—
—	—	—	—	—	—	—	—	—	—	—	—	—	—	—	—
—	—	—	—	—	—	—	—	—	—	—	—	—	—	—	—
Total	—	—	—	—	—	—	—	—	—	—	—	—	—	—	—

LD_{50} = ... ± ...... with range from ...... to ......

m = (log LD_{95}) = ... ± ...... with range from ...... to ......

LD_{95} = ... ± ...... with range from ...... to ......

If the tabulated value of X2 is more than the calculated value, the data are considered homogenous indicating thereby that there is a satisfactory agreement between the treatment and the data. If it is vice versa, the data is considered heterogeneous which may arise either because the individual test pest do not react independently to the toxicant or because the straight line does not satisfactorily describe the relation between the dosage and the probit (Jaglan, 2003).

Occasionally, negative X2 values are obtained which can be overcome by more decimal places for the values in the equation or by shifting the provisional regression hive around its LD_{50} axis in a counter clockwise manner.

In order to determine the relative performances, the toxicants have to be tested for a specific common character-toxicity. Unless, the toxicity is unified, the task of studying the relative toxicity will remain unresolved. Some of the common units used by toxicologists are LD_{50} (an amount of poison per unit weight of organism, which causes fifty per cent kill), LC_{50} (lethal concentration of the toxicant in the external environment of organism which kills 50 per cent of population), LT_{50} (the time required to kill 50 per cent of test population), KD_{50}, KC_{50}, KT_{50} (used where chemicals don't kill the pest but paralyse it-knock down effect), ED_{50}, EC_{50}, ET_{50} (where toxicant is used to sterilize them: effect is measured on fecundity).

Another purpose of employing the suitable bioassay method is to ensure good exposure of acaricide to mites quite close to that may occur in its natural habitat. Other important factors, which may influence the result, are the stage, age, sex, size of mite, temperature, relative humidity, food and handling of organism during testing. Accidental contamination during handling of mite and container will also affect the sensitivity of the method.

PART III:
Acarine as Biocontrol Agents

Bioagents are host specific, ecofriendly, environmentally safer and also cost effective as compared to pesticides, which leads to pest outbreaks, pest resurgence, and resistance in pest. Besides, pesticides disrupt biotic balance and ecosystem. Hence alternative systems of pest management are being developed of which biocontrol is an important component. While it is advantageous for any biocontrol agent to be host specific, the presence of other mites/ insects as alternative prey for natural enemies has been necessary to make biological control self–perpetuating (Van Emden, 1989).

Recognition of Acarines as effective biocontrol agents of insects, nematodes and weeds placed them as effective components of agricultural systems. Most of the work has still at exploratory stage but beginning has been made in this direction with few species being commercially mass reared to control thrips, stored insects and plant parasitic nematodes. Most of these mites show distinct preferences for prey of one group; like, *Bdellodes* spp. mainly attacks collembolans, *Blattisocius* sp. prefers stored product moths, *Hemisarcoptes* sp. are restricted to scale insects and *Macrocheles* feed on nematodes or housefly eggs. However, most of these can survive in the field on alternate diets in absence of preferred diet.

Acarine biocontrol agents usually are voracious by nature and possess well-developed searching and dispersal mechanisms. Many are fast runners and can be transported by wind, others locate their prey by kairomones and some utilize their hosts for dispersal. Another important criterion, which increases their chances for inclusion in successful biocontrol programme, is their greater reproductive potential than that of their pests. Species of *Blattisocius, Macrocheles, Acarophenax* and *Hemisarcoptes* reproduce more rapidly than their prey.

The food and feeding behaviour of mites depends on the type of mouth parts (esp. chelicerae), size of the prey (smaller consumed more than larger ones) and size of the predator. The significance of sheer size of acarine predator as compared to prey is also considered a promising character in pest control. Very small acarines (*Pyemotes* sp.) can subdue their prey by injecting them with quick action paralyzing toxin. Many acarine agents are quite sensitive to prey density and do best when pest populations are strongly clumped whereas as others showed enhanced feeding activity with increased pest population.

Some remarks on egg laying behaviour of predatory mites remains to be added which increased their survival and less energy is spent by hatched larvae in searching food. Predatory mites may lay their eggs in protective locations (eg. phytoseiids) or where the hatched larvae (eg. *Pergamasus* sp.) will have easy access to the prey. Brooding of clustered eggs occurs in sheltered nests of some cheyletid mites but if the hatchlings remain in the area too long, they may be eaten by the moth. Some parasitic mites (*Dicrochelus* sp.) use host tissue for oviposition.

The performance of natural enemies is also influenced by host plant characteristics (Lester *et al.*, 2000). Leaf surface texture and vestiture affect the performance of acarine predators associated with plants. Leaf domatia, minute cavities frequently located at the junctions of primary and secondary veins on the underside of woody dicot leaves often harbour various predaceous mites (Pemberton and Turner, 1989; Karban *et al.*, 1995). Phytoseiids show higher efficacy on hairy leaves, leaves with more pronounced veins and fruits with rough spurs which provide them overwintering shelter and better protection from their enemies. Anystids on the other hand provide better control on puberscent or glabrous leaves. In addition, many plants provide floral as well as extra floral nectars, pollen and even nutrients from other plant feeders (*i.e.* homopteran honeydew), natural foods known to be of importance for many generalist predators (McMurtry and Rodriguez, 1987). The quality and quantity of such natural foods differs from plant to plant with oblivious effects on predators. It was demonstrated that a stigmaeid ingesting palm date pollen will produce more than twice as many progeny as one feeding on castor bean pollen.

Following chapters will provide an insight into the work done on different aspects of acarine biocontrol agents.

Chapter 14

Acarines for Insect Control

The co-existence of mites and insects in a variety of habitats not only provides the opportunity for predation between these groups but also affords ideal conditions for the development of more complex associations such as commensalisms and parasitism. Among these associations, parasitism and predation by acarines is well understood, however, information is lacking on commensals and phoretic species. In parasitic mites, structural and biological adaptations are evident. In some ectoparasitic mites, only larval stage is parasitic (water mites and earth mites) but in others, all postembryonic stages are parasites on host (*Pyemotes* sp., *Podapolipus* sp.). The life cycle shows reduction in number of active developmental stages eg. in Erythraeidae and Trombidiidae only one active nymphal instar is produced while in *Pyemotes* sp., the female gives birth to mature adults. In parasitic mites, the chelicerae are needle like or hooked shape. In latter case, a feeding tube is produced by mite, which enters the host skin. Acarine predators generally restrict their feeding to the eggs or the immature stages of insects.

The first report dates back to 1868 when Shimer (1868) recorded *Hemisarcoptes malus* feeding on Apple bark louse, *Lepidosaphes ulmi* (L), which was later on, transferred from eastern to western Canada in 1917 to control the same pest. At that time it was rated as a successful biological control project. Riley (1874) found *Tyroglyphus phylloxerae* predating on major root infesting pest of grapes, *Daktulosphaira vitifollii* in France. *Pyemotes* was recognized in the 1880s as a natural enemy of many insect pests. One species was successfully mass reared on cotton boll weevil in Mexico (Rangel, 1901). A further Australian discovery of *Bdellodes lapidaria* as predator of collembolan, *Sminthurus viridis*, a major pest of pastures (Womersley, 1933) stimulated interest in acarine biocontrol agents. It was later introduced to South Africa to exert control on *Bourletiella arvalis*, another collembolan pest.

Table 14.1 showing acarines used for insects control is provided for ready reference.

Acaridae

Acarids are although pests of stored grains but could be used for insect and nematode control. *Tyrophagus putrescentiae* was observed to feed on corn rootworm (*Diabrotica undecimpunctata howardi;* Coleoptera) and grape phylloxera, *Daktulosphaira vitifollii* (Homoptera) eggs (Brust and House, 1988) and preferred them over fungi and dead arthropods.

Acarophenaxidae

Acarophenax lacunatus is another important parasite of *Rhizopertha dominica* (Matioli, 1997), *Cryptolestes ferrugineus* (rusty grain beetle) and *A. manhunkai* on *Alphitobius diaperinus* (lesser meal worm) (Sternkraus and Cross, 1993). *A. manhunkai* has short life cycle, high fertility and no evidence of harmful effects on humans and higher animals. The mite also exhibits female physogastry and phoresy. Phoretic mite leave the host body, parasitize its eggs and feeds on egg contents. A female produces seventeen off springs. Twenty-four hours after emergence, female undergo physogastry. Numerous mites develop with in the female body, reach adulthood and mate within themselves. After release from mother's body females track host whereas both male progeny and mother dies in this process. Four individuals/ 500ml jar containing 50 adults of pests, results in complete suppression of pest population after 45 days (Faroni *et al.*, 2001). Reduction in loss of stored wheat was estimated as 15-25 per cent.

A. tribolii parasitizes the grain beetle, *Tribolium* species. These are found where the host cuticle is thin, eg., on the intersegmental membranes and beneath the flight wings. After feeding on haemolymph for few days the young female leaves the host and feeds on beetle eggs. By this time, gravid females body is uniformly distended, on the third day she dies and 4-14 young females emerge through the enlarged genital orifice.

Table 14.1: Acarine Families Associated with Insects

Acarine Family	*Insect Pests*	*Pest Order*
Acaridae	Corn rootworm, grape phylloxera	Coleoptera, Homoptera
Acarophenaxidae	Stored product insects	Coleoptera
Anystidae	Citrus thrips, grape leafhopper, pine needle aphid, soil insects	Thysanoptera, Homoptera, Collembola
Arrenuridae	Mosquitoes	Diptera
Ascidae	Stored product insects, mushroom fly, mosquitoes	Hemiptera, Coleoptera, Lepidoptera, Diptera
Bdellidae	Lucerne flea	Collembola
Camerobiidae	Armoured scale insects	Homoptera
Cheyletidae	Stored insects	Coleoptera
Cunaxidae	Book lice, scales	Psocoptera, Homoptera
Erythraeidae	Bugs, pear psylla, apple tortricid	Hemiptera, Lepidoptera
Eupalopsellidae	Scale insects	Homoptera
Hemisarcoptidae	Armoured scale insects	Homoptera
Hydryphantidae	Aquatic arthropods	Diptera
Laelapidae	Corn rootworm	Coleoptera
Macrochelidae	Flies	Diptera
Parasitidae	Garden symphylan, flies	Symphyla, Diptera
Phytoseiidae	Thrips, white flies, scale	Thysanoptera, Homoptera crawlers
Pionidae	Mosquitoes	Diptera
Podapolipidae	Locust, grasshoppers, beetles	Orthoptera, Coleoptera
Pterygosomatidae	Blood sucking bugs	Hemiptera
Pyemotidae	Pink boll worm, stored insects, thrips, fire ants	Coleoptera, Lepidoptera, Hymenoptera
Trombidiidae	Aphids	Homoptera
Tydeidae	Scale insects	Homoptera

Anystidae

These are large, reddish, soft bodied and fast runner mites whose tarsus is longer than its claw. *Anystis agilis* is an important predator of grape leafhopper, *Erythroneura elegantula* and woolly pine needle aphid, *Schizolachnus piniradiata*. Female consumes 5-6 nymphs of grape leafhopper per day and lays 33 eggs while feeding on its culture. The mite feeds on all instars and morphs of aphids on alfalfa (Frazer and Nelson, 1981). It prefers citrus thrips, *Scirtothrips citri* to other pests. But long generation time (two generations/ year) and cannibalistic habits of this mite are a major hindrance in its mass rearing.

Arrenuridae

Larvae of *Arrenurus* sp. are the most common parasites on mosquitoes. The larvae locate host pupae and parasitize the adult mosquitoes as they emerge. When mosquitoes land on water to oviposit, the mites detach and continue their development in water. Female of the species produce 55-88 eggs per year. (Stechmann, 1978). *A. madaraszi* caused high mortality in preoviposition *Aedes* and *Culex* females (Reisen and Mullen, 1978).

Ascidae

Blattisocius tarsalis was recorded from 18 insect hosts but it prefers eggs and first instar larvae of flour moth (*Ephestia* sp.) and *Tribolium* spp. The other important hosts are *Plodia interpunctella, Rhyzopertha dominica, Ips grandicollis, Sitotroga* sp. The characteristics like high potential rate of increase as compared to moth, highly mobile feeding stages, phoresy on host, high speed (52-60 cm distance in 5 min.; 3 m distance in 30 min.) make it an effective biocontrol agent against moth in many parts of the world. Weekly releases of mite population are recommended in stored grains. The mite destroyed 3-5 eggs per day whereas gravid female consumed 5-6 eggs per day. The total life span, oviposition and fecundity of the mite are 20 days, 55 eggs and 5 eggs per day, respectively. *Platyseius* and *Cheiroseius* sp. (Ascidae) feed on mosquito eggs and larvae (Smith, 1983) whereas *Arctoseius cetratus* can reduce the egg hatchability of *Lycoriella auripila*, a pest of cultivated mushroom by 85 per cent (Binns, 1973).

Bdellidae

Bdellids are red, brown or greenish mites with snout like mouth parts, long palpi and two prominent sensilla on prodorsum. They actively hunt for their prey, sometimes tethering them with silken lines. *Bdellodes lapidaria* is another mite, which actively hunts for its prey collembolan, *Sminthurus viridis, Bourletiella arvalis*. Both these are pest of pastures in Australia and South Africa. The presence of more than 20 mites/ sq.m early in winter prevented pest outbreaks in Australia. Later on it was introduced in South Africa where it became established and effectively controlled the above-mentioned. pest (Wallace and Walters, 1974).

Camerobiidae

It is a small family whose members have characteristic long (stilted) legs, weak palpi and looped peritremes. *Neophyllobius* sp. is known to feed on crawlers of California red scale, *Aonidiella aurantii* and European fruit scale, *Quadraspidiotus ostreaeformis*.

Cheyletidae

Acaropsis docta feeds on stored product insects like *T.granarium, S. granarium* (Kapil and Bhanot, 1971), *R. dominica, T. castaneum* and *Lasidoderma sericornae* (Al-Badry *et. al.*, 1980). A female deposits 80-121 eggs in 8 days (Al Badry *et al.*, 1980; Yousef *et al.*, 1992) either in clusters or scatters. As soon as the female lays her daily quota of eggs, she prepares a net of web around them. It requires 14 to 15 days to complete its development. Feeding potential of different stages are reported in literature (SomChaudhary and Mukherjee, 1971a; Kapil and Bhanot, 1973). The male and female consume an average of 10 and 20 eggs in 10 days. *Cheletogenes ornatus*, an orangish coloured mite, is a predator of first instar of armoured scale insects. It consumes 90 crawlers during adult life and produces about a dozen progeny. *A. docta, A. sollers* are reported to feeds on *Trogoderma* eggs and show preference for young eggs (1 day old). A mite consumes 1-2 eggs/day and 37-40 eggs in its life.

Cunaxidae

These are yellow red or brown in colour and are fast runners. Their palps are ornamented with strong spines. They hunt and fasten their prey with silken threads excreted through their mouth parts.

Cunaxa capreolus of Cunaxidae family is an important predator of booklice. Zaher *et al.* (1975) reared cultures of this mite successfully on booklice. At 30°C the mite completed a generation in about four weeks on booklice and each female deposited 45 eggs. One mite consumed 230 booklice during its development. Another cunaxid, *Cunaxoides parvus* is considered an important predator of the oystershell scale, *Lepidosaphes ulmi* on apples in North America (Schruft, 1971).

Erythraeidae

These are large, reddish mites with two pairs of prodorsal sensilla. The larvae are parasitic on other arthropods whereas nymphs and adults are predatory in nature. Two members of Erythraeidae, *Leptus* and *Bochartia* sp. are reported to feed on a mango pest, *Drosicha mangiferae*. They killed the pest by sucking their fluid. On one host, 1-60 mites were recorded by Tandon and Lal (1976) but no further information is available.

Eupalopsellidae

They are yellow to orange coloured mites with characteristic long palps and chelicerae. They are often associated with armoured scale insects. *Saniosulus nudus* completes its life cycle in three weeks when fed on Florida red scales, *Chrysomphalus aonidum*. Each female produced 40-50 eggs (Gerson and Blumberg, 1969) and consumed 259 prey eggs (Zaher *et al.*, 1984). Rao and Sankaran (1969) also recorded them feeding on sugar cane pest, *Melanaspis glomerata.*

Hemisarcoptidae

It is the only family, which is at present, is being exploited for armoured scale insects control in many parts of the world. These are whitish, soft-bodied mites having wide range of climatic tolerances. All species of genus *Hemisarcoptes* are feeders on armoured scale insects. Female mites feed on adult scales and their eggs and deposit their own eggs on the adult insects. Some immature mites continue to feed there while others move to additional hosts. *Hemisarcoptes coccophagus, H. cooremani* and *H. malus* are the major mortality factor of the oystersheel scale *Lepidosaphes ulmi* on apple, date palm scale *Parlatoria blanchardi* and citrus scale insects. Their successful introduction in New Zealand (1987-88) led to 80 per cent reduction in scale insect population (1991) in orchards. Mites disperse as deutonymphs, which lack mouthparts. In absence of host, they live for up to a fortnight. Dispersal is through lady beetles *Chilocorus, Lindorus* sp. Their association with beetle lasts for 5-20 days, during which they get some secretions from beetle body without damaging her. This association is termed as transitional between phoresy and parasitism. In absence of beetles, mites can track host scales. Mites cut through host shields by active slicing movements (Gerson and Schneider, 1981). Methods for mass rearing of *Hemisarcoptes* are available.

Hydryphantidae

Hydrophantids are normally found in ponds, springs and seepage areas. They are associated with mosquitoes (Mullen, 1975; 1977), blackflies, biting midges (Smith and Oliver, 1986) and horse flies (Smith, 1983). *Panisopsis* sp. is known to curtail the flight activity and host seeking of *Aedes ventrovittis* (Fenley, 1966). Some larvae attach to young emerging mosquitoes thus parasitizing both the sexes whereas others attack only gravid females (Mullen, 1977).

Laelapidae

Many species of this family are blood-sucking parasites of birds and mammals but others are nest inhabiting or free-living predators of small invertebrates. Chiang (1970) recorded that soil borne *Androlaelaps* and *Stratiolaelaps* spp. usually accounted for 20 per cent natural control of corn rootworms (*Diabrotica longicornis, D. virigifera*). A generation required 60 days at 20°C and 95 per cent relative humidity. Some adults lived upto 8-9 months and fecundity was 35 eggs/ female. In field experiments, manure application increased mite population three fold, which led to 63 per cent pest control. Another soil borne predator, *Hypoaspis* sp. is currently being exploited against fungus gnats, springtails, thrips pupae and plant parasitic nematodes. This tiny (0.5 mm) light-brown mite inhabits the top 1/2" layer of soil. The female *Hypoaspis* mites lay their eggs in the soil, which hatch in 1-2 days, and the nymphs and adults feed on the soil dwelling pests. Each *Hypoaspis* mite consumes 5-20 prey or eggs per day. It was also recorded to feed on thrips pupae, Collembolans and beetles and brought about 70 per cent reduction in *Oryctes* (beetle) egg hatch. Their entire life cycle was completed in 7-11 days. Commercially available cultures are shipped as adults, packed in a medium of bran, vermiculite and peat for easy dispersal with release rates as 10,000-25,000/per acre. Cultures can be kept for 2-3 weeks at 45-55°F, if not in immediate use.

Limnesiidae

Limnesia jamurensis (Limnesiidae) adult consumes seven and eight eggs of *Anopheles* and *Culex* larvae /day, respectively. When eggs of both are offered, consumption increases to 18 eggs /day. The attachment of limnesiid larvae to host's thorax damage host musculature, thus cause a reduction in flight activity (Smith, 1988).

Macrochelidae

These are fast moving, free living predators common in habitats, which are rich in decaying organic material including manure. Females are often disseminated by various flies and beetles. Several species of *Macrocheles* preferred eggs and immatures of housefly. *M. muscaedomesticae* completed its life cycle in 4-5 days and laid 160 eggs per female on housefly cultures with feeding rates upto 20 immatures per day. It gave effective control in cattle manure by reducing the house fly population by 90 per cent (Rodriguez *et al.*, 1970). It provided natural control of lesser house fly, *Fannia canicularis* in a small animal farm. Other macrochelid, *M. glaber* predates on Australian bush fly (*Musca vetustissima*) and almost total control was obtained when 50 mites were added in 1000ml dung pad with 300-400 fly eggs as no fly emerged (Wallace *et al.*, 1979). *M. peregrinus* (generation in 3 days) was introduced from Australia to South Africa to control bush fly and buffalo fly, (Wallace and Holm, 1983), provided good control initially and colonized in 180,000 Km^2 area around the release sites. However, later on no reduction in pest population was observed and failure was attributed to insufficient alternate prey.

Parasitidae

Parasitids are common predators in soil. They are often dispersed during their deutonymphal stage by various insects especially beetles and flies. *Pergamasus quisquiliarum* is closely associated with plant root system and could be an important factor in regulating the rate of symphylid population increase in the field. It feeds on garden symphylan, *Scutigerella immaculata*, a soil borne pest of many crops. The mite completed a generation in 17 days whereas the pest required 87 days to complete its life cycle. Female deposited 33 eggs near plant roots. During its life, the mite consumed an average of 15 symphlans, actively searching for its prey (Berry, 1973). Another parasitid, *Poecilochirus monospinosus* feeds on housefly immatures in poultry manure (Wise *et al.*, 1988). The life cycle is completed in 17 days and 89 eggs per female are laid. Deutonymphs destroy 5 eggs or 1st instar per day whereas females destroyed 13 eggs or 24 larvae per day. It provided an effective control in poultry manures against housefly. *Parasitus hituherosus*, a member of Parasitidae is considered a possible bio control agent of *Heteropeza pygmae* in mushroom compost (Al-Amidi and Downes, 1990).

Phytoseiidae

These are the best-known predators among the Acari and may easily be mass reared and shipped. These fast runner mites are cultured on acarid mites, spider mites or on alternate diets (pollen). *Amblyseius (Neoseiulus) cucumeris, A. barkeri* (Hansen, 1988), *Typhlodromus occidentalis* (Field, 1976) and *Euseius hibisci* (Tanigoshi *et al.*, 1984) are known to predate on different species of thrips. They are mostly generalist predators who show lower rate of increase on pest mite than on other food, more stable populations and do not disperse in absence of prey. IPM technology is developed for cucumber and pepper using *A. (N.) cucumeris* culture (Jacobson *et al.*, 2001). Initially, Sprinkling Method (loose bran based medium) was used which needed more number of biocontrol agent as some material dropped from the leaves and repeated sprinkling was required. This method provided success in pepper but not in cucumber. Then, it was replaced by plastic tubes containing phytoseiid mites and acarid mites as their diet, which were hung on plants. Soon, plastic tubes were replaced by paper sachets, which contained special food formulations for prey mite. One sachet every third plant was released every third week to get effective control of thrips. Due to low moisture, this method caused mortality of pest mite used for predator culture. Another major drawback was that cucumber plants do not come in contact with each other till third week of planting, by that time thrips completed one generation on the two third unprotected plants. Ordinary paper was then, replaced by waxed paper sachets of specific porocity which retained moisture as well as allow the passage of air. Diapause resistant strain of phytoseiid was used in this culture and one sachet/ plant was released at the time of planting so that phytoseiid mites multiply in time and was there when thrips attack came. This method provided complete suppression of thrips population and culture was active for nine weeks.

A. cucumeris eggs mature to adult in 6-9 days and its first and second stage nymphs and adults are predacious. Adults are beige-pink in color and live for 20 days. Cultures are available and they are shipped as adults in a bran formulation. Beneficial mites should be placed at the base of plants to ensure maximum efficiency. Successful control of thrips was achieved in Brazil and Egypt.

Pionidae

The legs of pionids have swimming hairs and tarsal claws. They have a weak and smooth integument. The nymphs and adults of *Piona nodata* are voracious predators of mosquito larvae (Smith, 1983). Consumption of host egg is 100–120 eggs /female. Mite nymphs destroy 5 first instar/ day whereas female consumes 20-25 fourth instar/ day.

Podapolipidae

Podapolipid mites are small mites with one to three pairs of legs. Their body shape varies from vermiform to pear shaped to sac like body. They undergo a shortened life cycle. The sexually mature males hatch from eggs and mate with larviform females, which subsequently molt. *Locustacarus* sp. parasitizes the tracheae of acridids and their feeding is found to be detrimental to host health. Another species, *Podapolipoides grassi* is a common, most wide spread ectoparasite of locusts. It infests third instar nymphs and subsequent stages and normally occurs on thorax or anterior abdominal segment and at the base of wings. The parasite spends its entire life cycle on the locust, dispersing during host mating. Their feeding leads to decrease in host vigour (Gauchat, 1972).

Pterygosomatidae

This acarine family is associated with blood sucking bugs and cockroaches. These red, small to medium sized mites are parasitic in nature. *Pimeliaphilus plumifer* parasitizes blood-sucking bugs of the family Triatomidae. At 30°C and 15-20 per cent relative humidity, complete their life cycle in 60 days and live for 54 to 78 days. On an average 60 eggs are laid by female. When present in large numbers (25 mites/ insect), they kill their host with in one hour. These are not very host specific and can survive on any triatomine bugs within their range (Anderson, 1968). Death of the host is due to injected toxin. Furthermore, egg capsules produced by parasitized cockroaches fail to hatch; hence host population is reduced considerably. Mites also serve as vectors for pathogenic virus and bacteria which enhances the possible use of this group for the biological control of blood sucking bugs.

Pyemotidae

Pyemotid body is milky white and spindle or round shape. *Pyemotes tritici,* the straw itch mite is also an effective biocontrol agent of insect pests. It attacks all stages of *Plodia interpunctella, Cadra cautella* (Lepidoptera), *Oryzaephilus mercati, Lasioderma serricorne* (Coleoptera) (Bruce and LeCato, 1979; 1980). It is easy to rear (synthetic diet is available: Bruce, 1989), population can be synchronized with pest population, high reproductive potential (100-200 progeny/ female), short life cycle (7 days), female make up 98 per cent population and seek out host immediately upon emergence and show phoresy, physogastry behaviour. It inject toxin, paralyzes the host before killing. Till mid 80's not much progress was made regarding its exploitation as biocontrol agent because it causes dermatitis in humans. Moser *et al.* (1987) observed two groups of *Pyemotes* spp., one was parasitic on Scolytid beetle and venom less while the other group was polyphagous and venomous. The latter group was found to be very lethal to insects but show reduced dermatological risks to humans. *P. tritici* is also used as biocontrol agent against red fire ant, *Solenopsis invicta.*

P. herfsi parasitizes *Tribolium confusum* larvae and can locate the beetle 7 cm beneath of bran. The progeny of a single mite kills all larvae of a single beetle with in ten days. It also parasitizes over wintering caterpillars of pink bollworm, *Pectinophora gossypiella* (85 per cent mortality). Mite induced mortality reached upto 85 per cent in the first generation larvae of pink bollworm, an important pest of cotton. It was found to parasitize the soft scale *Saissetia oleae,* a citrus pest, in Argentina (Mallea *et al.,* 1983).

Adactylidium sp. is another mite of this family which parasitizes the eggs of thrip, *Gynaikothrips ficorum,* a pest of ornamentals in Egypt and Brazil. It is quite abundant in nature and is associated with all stages of thrip which is a contributory factor for thrip control.

Trombidiidae

Tombidiid mites are large, usually red velvet mites whose larvae are parasites of insects whereas; their nymphs and adults are predatory in nature. Mites preferred the ventral sides of small and medium sized hosts whereas lateral sides of large sized hosts were preferred. *Allothrombium pulvinum* larvae are ectoparasites on aphids whereas nymphs and adults predates upon aphids and lepidopterous eggs. Two or more larvae per aphid kill the host in 1-3 days (Zhang and Xin, 1989; Dong *et al.*, 1996). *Dinothrombium pandorae* larvae parasitize grasshoppers, however, its nymphs and adults feed on termites. One female lays approximately 80,000 eggs during her lifetime (Newell, 1979). *Eutrombidium locustarum* larvae parasitize first instar to adult stage of host at the rate of 35-mites/host. More than 75 per cent parasitization is observed in grasshopper population. Mite feeding interferes with normal adult flight by causing grasshopper wings to become battered and even broken. Mite nymphs and adults seek out host eggs in the soil and devour them; a mite consumes about five eggs/day. Each female lays about 4700 eggs and each generation requires 60 days to complete in the laboratory.

Tydeidae

Several members of Tydeidae family are often found in association with soft scales, feeding on their honeydew. They are beneficial in the sense that they eliminate the honey dew secreted by scales on which sooty mould fungi develops, thereby reducing the damage caused by mould.

Chapter 15

Acarines for Nematode Control

The acarines predating upon nematodes are euedaphic forms with rapid generation time and high reproductive out put. They do possess certain survival strategies too Nematodes sometimes provide a high nutrient rich diet to nematophagous acarines, which develop faster and reproduce more under such conditions. In *Macrocheles muscaedomesticae* 23.11 ±5.58 progeny per female per day is produced in presence of both, nematodes and house fly eggs (Rodriguez *et al.*, 1962). *Gamasellodes vermivorax* completes its life cycle on rhabditid nematode just in 10 days at 24 °C and population increases four fold after 16 days depending upon nematode densities in microcosms (Walter, 1987). In greenhouse cultures of *Tylenchulus semipenetrans* and root–knot nematode at Florida, USA, the densities of nematophagous mites often exceeded 1000 mites per liter (Walter *et al.*, 1993).

Soil acarines are instrumental in keeping the populations of plant parasitic nematodes under check in some of the undisturbed ecosystems. Acarines belonging to all the four sub-orders *viz.*, Astigmata, Cryptostigmata, Mesostigmata and Prostigmata have shown varying degrees of nematophagous activity (Table15.1). The most of the nematode predacious acarines belong to Mesostigmata group and are quite abundant and frequent geographically in the soil ecosystem. Mites belonging to these suborders are discussed briefly, which are associated with nematodes.

Table 15.1 Mites associated with nematodes

Mites	*Nematodes*
I. Astigmata	
Caloglyphus sp.	*Heterodera glycines, Meloidogyne incognita*
Rhizolyphus echinopus	*A.ritzemabosi, Ditylenchus dipsaci, D. destructor, H. avenae, Longidorus elongatus, M. javanica, Xiphinema index*
Sancassania ultima	*Globodera rostochiensis, H. schachtii, M. hapla, M. incognita, M. javanica, Paratylenchus* sp., *Pratylenchus* sp., *Tylenchorhynchus* sp.
Schwiebea rockettii	*Radopholus citrophilus*
Tyrophagus putrescentiae	*Acrobeles* sp., *Acrobeloides* sp., *Anguina tritici, Aphelenchoides composticola, Cephalobus* sp., *Hirschmanniella oryzae, Heterorhabaditis heliothidis, Longidorus pisi, M. incognita, M. javanica, Steinernema feltiae, Rhabditis* sp. *Tylenchorhynchus mashhoodi*
T. similis, T. zachvatkini	*Acrobeloides* sp., *Aphelenchoides* sp., *Aphelenchus* sp., *A. avenae, Panagrolaimus* sp.
II. Cryptostigmata	
Ceratozetes sp., *Fuscozetes* sp., *Galumna* sp., *Haplozetes* sp., *Nothrus* sp., *Oppia* sp., *Peloribates* sp., *Pilogalumna* sp.	*Acrobeloides* sp., *Aphelenchoides* sp., *Aphelenchus* sp., *A. avenae, Panagrolaimus* sp.
Opiella nova	*Pelodera lambdiensis, Tylenchorhynchus martini*
Pergalumna sp.	*Acrobeloides* sp., *Aphelenchoides* sp., *Aphelenchus* sp., *A. avenae, Panagrolaimus* sp., *Chiloplacus* sp., *Rhabditis* sp.
P. omniphagus	*P. lambdiensis*

Contd...

Table 15.1–Contd...

Mites	*Nematodes*
III. Mesostigmata	
Alliphis halleri, Antennoseius sp.	*Acrobeloides* sp., *Aphelenchoides* sp., *Aphelenchus* sp., *A. avenae, H. rostochiensis, Panagrolaimus* sp.
Arctoseius cetratus	*Acrobeloides* sp., *Aphelenchoides* sp., *Aphelenchus* sp., *A. avenae, Panagrolaimus* sp.
Asca garmani, A. nesoica, A. piloja	*Acrobeloides* sp., *Aphelenchoides* sp., *Aphelenchus* sp., *A. avenae, Panagrolaimus* sp.
Cosmolaelaps sp., *C. vacua*	*Acrobeloides* sp., *Aphelenchoides* sp., *Aphelenchus* sp., *A. avenae, Panagrolaimus* sp.
C. claviger	*Rhabditis* sp.
Crassicheles sp.	*Acrobeloides* sp., *Aphelenchoides* sp., *Aphelenchus* sp., *A. avenae, Panagrolaimus* sp.
Dendrolaelaps latior, D. procornutus, D. strenzkei, D. zwoelferi	*Acrobeloides* sp.
Euandrolaelaps karawaiewi	*Acrobeloides* sp.
Eugamasus sp., *Eviphis* sp.	*Steinernema feltiae*
Gamasellodes vermivorax, G. bicolor	*Acrobeloides* sp., *Aphelenchoides* sp., *Aphelenchus* sp., *A. avenae, Deladenus durus, Chiloplacus propinquus, Mesodorylaimus* sp., *Pelodera* sp., *Rhabditis terricola, S. feltiae, Panagrolaimus* sp.
Gamasellus montanus	*Acrobeloides* sp., *Aphelenchoides* sp., *Aphelenchus* sp., *A. avenae, Deladenus durus, Chiloplacus propinquus, Mesodorylaimus* sp., *Pelodera* sp., *Rhabditis terricola, S. feltiae, Panagrolaimus* sp.

Contd...

Table 15.1–Contd...

Mites	*Nematodes*
Hypoaspis aculeifer	*Heterodera schachtii, M. chitwoodi, M. hapla*
H. calcuttaensis	*Acrobeles* sp., *Acrobeloides* sp., *A. tritici, Aphelenchoides composticola, Aphelenchus avenae, Aporcelaimellus nivalis, Aquatides thornei, Basiria* sp., *Cephalobus* sp., *Chiloplacus symmetricus Dorylaimus stagnalis, Discolaimus silvicolus, Hemicriconemoides mangiferae, Hemicycliophora dhirendri, Heterodera mothi, H. oryzae, Helicotylenchus indicus, Hoplolaimus indicus, Longidorus* sp., *M. incognita, M. javanica, Mesorhabditis* sp., *Mononchoides fortidens, M. longicaudatus, Mononchus aquaticus, Mylonchulus dentatus, Paralongidorus citri, Rhabditis* sp., *Paratrichodorus* sp., *Scutellonema* sp., *Tobrilus* sp., *Tylenchorhynchus mashhoodi, Xiphinema basiri*
Janetiella pyriformis	*Ditylenchus myceliophagus, Heterorhabditis heliothidis*
Laelaspis sp.	*D. myceliophagus, H. heliothidis*
Lasioseius sp.	*A.tritici, A. composticola, M. incognita, Longidorus pisi*
L. athiasae	*Meloidogyne* sp.
L. penicilliger, L. scapulatus	*Tylenchorhynchus dubius, Paratylenchus* sp., *Pratylenchus* sp., *A. avenae, Acorbeloides* sp., *Caenorhabditis* sp., *Cephalobus* sp., *Diplenteron* sp., *M. incognita, Mononchus* sp., *Paraphelenchus* sp., *Seinura* sp., *Rotylenchus* sp.
L. thermophilus	*D. myceliophagous*

Contd...

Table 15.1–Contd...

Mites	Nematodes
L. youcefi	*A. avenae, Acorbeloides* sp., *Caenorhabditis* sp., *Cephalobus* sp., *Diplenteron* sp., *M. incognita, Mononchus* sp., *Paraphelenchus* sp., *Seinura* sp.
Macrocheles sp.	*Acrobeloides* sp., *Aphelenchus* sp., *Rhabditis* sp., *Panagrolaimus* sp., *S. feltiae*
M. muscaedomesticae, M. insignitus, M. schaeferi	*A. avenae, A. composticola, Diplogaster* sp., *Panagrolaimus* sp., *Rhabditella leptura, Rhabditis elongata, R. teres, Panagrellus redivivus, Tylenchulus semipenetrans*
Mixozercon sp.	*A. avenae, A. composticola, Diplogaser* sp., *Panagrolaimus* sp., *Rhabditella leptura, Rhabditis elongata, R. teres, Panagrellus redivivus, T. semipenetrans*
Protogamasellus hibernicus, P. mica, Paragamasus sp., *Pergamasus crassipes, Pseudoparasitus* sp., *P. austriacus, Proctolaelaps juradeus, Ololaelaps veneta,*	*Acrobeloides* sp.
Rhodacarellus silesiacus, R. subterraneus	*Acrobeloides* sp.
Rhodacarus denticulatus	*Acrobeloides* sp.
R. roseus, Veigaiia sp., Zercon sp., Skeironozercon tricazus	*Paratylenchus* sp., *Pratylenchus* sp., *Rotylenchus* sp., *T. dubius*

Contd...

Table 15.1–Contd...

Mites	*Nematodes*
IV. Prostigmata	
Alicorhagia fragilis	*Acrobeloides labiatus, A. avenae, H. heliothidis, Panagrolaimus* sp., *S. feltiae*
Alycus roseus, A. usitata, Eupodes sp., Linopodes sp.	*H. heliothidis, S. feltiae*
Neocunaxoides andrei	*Panagrolaimus rigidus*
Paratydens sp.	*P. rigidus*

Linford and Oliveira (1938) observed 52 enemies of the nematode species. Out of which six mite species (unidentified) were recognised as nematophagous. It was the first record of mites, which fed on nematodes. They indicated the need for new techniques and ways, which could stimulate the predatory activities of mites against plant parasitic nematodes. However, no attention was paid to these nematophagous mites for almost two decades. Few reviews are available on these mites (Walia, 2003; Gulati *et al.*, 2005). Nematophagous predatory mites showed mean abundance of 37500 top to 370000 per sq m down to a depth of 15 cm in a range of different soils (Karg, 1983).

Astigmata

These are whitish slow moving mites whose empodial claws are well developed and carry nude or slightly barbed opisthomal setae. These mites mainly occur in stored foods but few species occur in soil also which feeds on nematodes. *Tyrophagus putrescentiae, Caloglyphus* sp. and *Rhizoglyphus echinopus* was recorded as feeding on several plant parasitic nematodes (Bilgrami and Tahseen, 1992; Bilgrami, 1994; Walia and Mathur, 1995; Ranjana *et al.*, 2001). Small nematodes were completely devoured, whilst larger ones were cut into pieces and sucked (Sturhan and Hampel, 1977; Walter *et al.*, 1986; Sell, 1988; Walia and Mathur, 1995). *T. putrescentiae, T. similis* and *T. zachvatkini* could be cultured on a nematode diet in addition to their regular diet. There are few reports which suggest preference for nematode diet over other diets (Walter and Ikonen, 1989). *T. putrescentiae* also multiplied on root-knot nematode, *Meloidogyne javanica* and seedgall nematode, *Anguina tritici*, successfully (Ranjana, 2002). The population of the nematode decreased in the presence of the mite. The mites helped in mitigating the suppressive effect of nematode on plant growth (Salem and Abo–Korah, 1981). *R. echinopus* consumed a variety of nematodes including infective stages of entomopathogenic nematodes, *Heterorhabditis heliothidis* and *Steinernema feltiae* (Epsky *et al.*, 1988). *Schwiebea rockettii* fed on a variety of lower organisms including dead nematodes and could be cultured on a plant parasitic nematode, *Radopholus citrophilus* (Walter and Kaplan, 1990).

The size of the prey (Ranjana *et al.*, 2003), prey and predator density (Ranjana *et al.*, 2005) also influenced the predatory potential of acarid mites. Consumption of small sized nematodes (*M.incognita*,

0.3 mm) by mites was maximum (50 per cent) followed by medium sized nematodes (*A. tritici,* 0.7 mm) (59 per cent) and large sized nematodes (*Longidorus pisi,* 2-4mm) (9 per cent) in 24 hrs. Prey and predator density exhibited a positive response on predatory potential of nematophagous mites.

Cryptosigmata

There are limited reports of predation by oribatid mites on nematodes as they are heavily sclerotised and slow moving in nature. Rockett and Woodring (1966) observed an oribatid mite species, *Pergalumna omniphagus* predating upon free-living nematode, *Pelodera lambdiensis*. This was a significant finding for these primarily detritus feeders. *Pergalumna* sp. voraciously fed upon *Chiloplacus* sp. and *Rhabditis* sp., the two free living nematode genera. Specimens of *Galumna* sp., *Fuscozetes* sp. and *Nothrus* sp. fed less frequently whereas *Ceratozetes* sp. only chewed and did not devour the nematodes.

Further observations on feeding behaviour indicated that the mite neither searched nor preferred nematodes and the time required for a tritonymph or adult pergalumnid to grasp and completely devour an average sized nematode (± 500µm) varied considerably from 15 seconds to 25 minutes. As many as three nematodes could be consumed successively. The work of Muraoka and Ishibashi (1976) added a list of 22 cryptostigmatid mite species to our knowledge of nematophagous mites. Nematodes predator by oribated are also documented by Rockett (1980) who observed individual mites of *Pergalumna* spp. feeding on plant parasitic nematodes.

Mesostigmata

The most of the nematode predacious acarines belong to this group and are quite abundant and frequent geographically in the soil ecosystem. Ascid mites are diverse and abundant in many soils where they are voracious predators of small invertebrates (Walter and Ikonen, 1989). They exhibit a variety of reproductive modes from obligate bisexuality to obligate parthenogenesis. They require decaying fungal sporocarps for their dispersal, which is necessary to exploit patchy and ephemeral resources. Walter and Lindquist (1995) further recorded that all female ascid species rarely form the phoretic associations with insects. *Lasioseius* species reduced the seedgall nematode population by more than 82 per cent in *in vitro* trials with in 72 h of its inoculation (Ranjana, 2002). It was successfully multiplied on *M. incognita,* fungivorous nematode, *Aphelenchoides composticola* and *A. tritici* (Ranjana, 2002). *L. penicilliger, L. scapulatus* have also been reported to feed upon nematodes, *Tylenchorhynchus dubius, Paratylenchus* sp., *Pratylenchus* sp., *Rotylenchus* sp., *Cephalobus* sp., *H. glycines, Meloidogyne* spp., *A. avenae, Acrobeloides* sp., *Caenorhabditis* sp., *Seinura* spp. etc. (Sharma, 1971; Van de Bund, 1972; Muraoka and Ishibashi, 1976; Imbriani and Mankau, 1983; Inserra and Davis, 1983; Ishibashi *et al.*, 1987). *L. thermophilus* is found associated with mushroom compost and feeds on the eggs of *Ditylenchus myceliophagus*. The mite population increased from 10 and 50 to 603 and 1611, respectively in the presence of nematodes whereas nematode population declined to 18,550 (50 mites inoculated) from 504,000 in control (absence of mites). Its possible use as a biocontrol agent was also suggested by Cayrol (1970) against *D. myceliophagus* in mushroom beds.

Thirteen nematode species belonging to different trophic groups were fed upon by *H. calcuttaensis* in *in vitro* studies (Bilgrami, 1994). *Hypoaspis miles* are tiny (0.5 mm) light-brown mites that naturally inhabit the top 1/2" layer of soil where fungus gnats, as well as springtails and thrips pupae dwell. The female *Hypoaspis* mites lay their eggs in the soil, which hatch in 1-2 days, and the nymphs and adults feed on the soil-dwelling pests. Each *Hypoaspis* mite will consume 5-20 prey or eggs per day. Under certain field ecosystems the mesostigmatic mite, *H. aculeifer* could limit the populations of *M. chitwoodi, M. hapla* and *Heterodera schachtii* (Inserra and Davis, 1983). Addition of organic manures, leaf litter, intercropping in field crops favours the development and multiplication of the mites which in turn helps in regulating the population of plant parasitic nematodes (Karg, 1983).

The food and feeding behaviour of mites depends on the type of mouth parts (esp. chelicerae), size of the prey (smaller consumed more than larger ones) and size of the predator (large predators survive on both nematode diet and arthropod diet). Nematodes sometimes provide a high nutrient rich diet to nematophagous acarines, which develop faster and reproduce more under such conditions. *Gamasellodes vermivorax* completes its life cycle on rhabditid nematode just in 10 days at 24°C and population increases four fold after 16 days depending upon nematode densities in microcosms (Walter, 1987). *Contortylenchus brevicomi*, a nematode found in galleries of bark beetles is essential for the development of *Dendrolaelaps neodisetus*, which completes its life cycle in 6 to 13 days at 25°C on the nematode diet (Kinn, 1984). *G. vermivorax*, *Macrocheles* sp. and *Cosmolaelaps vacua* take lesser time to develop on a nematode diet than on an arthropod diet and lay eggs only in presence of nematode diet (Walter *et al.*, 1987). *M. muscaedomesticae* nymphal stages (Rodriguez *et al.*, 1962) prefer nematodes to their regular diet. Fungus feeding and free-living nematodes, *Aphelenchus avenae, Aphelenchoides composticola, Panagrellus redivivus, Rhabditis elongata* and *R. teres* also provide a rich food resource for *M. muscaedomesticae* (Ito, 1971). All the stages of small nematodes and immature stages of large nematodes were predated upon effectively by *Rhodacarellus silesiacus, R. calcarulatus* and *Rhodacarus roseus* (Karg, 1961; 1971).

In greenhouse trials, the citrus nematode, *Tylenchulus semipenetrans* could be restricted by *Macrocheles muscaedomesticae* successfully (Osman *et a1*.. 1988). The mesostigmatic mites have been judged as better predators in soil ecosystem than astigmatids which are rare in soil (Walia and Mathur. 1996). These also hold potential to be used for the nematode pests of mushroom. The preliminary experiments with *H. ca1cuttaensis* against *A. composticola* (Walia and Mathur, 1994) and *L. thermophilus* against *D. myceliophagus* are encouraging (Cayrol. 1970). *H. ca1cuttaensis, L. athiasae* and *Protogamasellus discorus* have been found to reduce soil population and root galls caused by root-knot nematodes. The mite population increased eight times after 45 days in the presence of nematodes (Walia and Mathur. 1996; Amin *et al.*, 1999).

Prostigmata

Among prostigmatids, the two genera, *Alicorhagia* and *Alycus* are important nematode predators (Walter 1988 a, b). Nematodes though consumed whole or in bits are difficult to recognize in gut contents owing to lack of skeletal structures except stylet, which is too small to locate easily. *Alycus roseus* is a nematode specialist whereas *Alicorhagia fragilis* though an efficient nematode predator consumed other food too. The infective stages of the two entomopathogenic nematodes, *H. heliothidis* and *S. feltiae* are reported to be predated upon by these mite species (Epsky *et al.*, 1988) and do hamper the populations of these nematodes in soil. *Neocunaxoides andrei* though develop faster on a nematode diet than an arthropod diet yet females live longer on arthropod diet compared to nematode diet (Taha *et al.*, 1988). Another cunaxid mite, *Coleoscirus simplex* reported from greenhouse cultures of root-knot nematode in Florida could develop and reproduce on second stage juveniles of the nematode and not the adults and egg masses (Walter and Kaplan, 1991).

The predacious mites are important in breakdown of litter and decomposition process. The bacteriophagous nematodes are predated upon by the mites, which indirectly enhance the process of decomposition in presence of decomposing bacteria, which are otherwise grazed upon by the nematodes in the absence of predatory acarines (Santos and Whitford, 1981; Santos *et al.*, 1981; Elkin and Whitford, 1982; Laakso and Setala, 1999).

Predatory acarines play very different roles in insect–nematode interrelationship in woody trees. The mites may feed on *Bursaphelenchus xylophilus*, the nematode involved in pine-wilt disease of pines in Japan (Tamura and Enda, 1980 a, b) or the nymphal stages feed on nematodes which increase the population of adults instrumental in lowering the population of bark beetles (*Ips* sp.) by predation (Kinn, 1984) or the mite may hamper the control of insects by feeding on entomogenous nematodes (Kinn, 1983; Epsky *et al.*, 1988).

Sharma (1971) and Mankau (1980) while reviewing the control of nematode pests by natural enemies, believed that mites, which have greater searching ability than other predators, should be further investigated, despite some drawbacks, such as non-specificity, being limited to the upper soil layer, which is only a portion of the vertical range of some pest nematodes, restriction by size to soils with pore spaces which allow for their passage and initiation of attack only by fortuitously coming upon the nematodes. Few encouraging reports suggest that under certain conditions predatory mites could protect field crops from nematode damage. It is quite evident that appropriate field experiments should be done to evaluate the significance of various mites against pestiferous nematodes.

Chapter 16

Acarines for Weed Control

Acarines feeds on weeds or such plants which successfully repress or compete with agricultural crops causing yield reductions and responsible for water clogging. Harris (1973) and Goeden (1983) proposed scoring systems for predicting the efficiency of arthropods introduced for weed control shortly before, during and after their initial release. Few acarine families are associated with weed control, which is summarized below with the help of Table 16.1.

Tetranychidae

Spider mites are recognized by their needle like chelicerae and a thumb claw complex. These mites produce webbing and are important plant pests, which suggested their use for weed control (Andres, 1983; Cromroy, 1983). *Tetranychus desertorum* effectively reduced the cactus prickly pear, *Opuntia* spp. by 75 per cent in two years of its introduction (Hill and Stone, 1985). Atleast twelve species of polyphagous tetranychids are suggested for waterhyacinth control. The strongly webbing colonies are observed to move slowly along growing shoots for feeding and ovipositing. The webbing can envelop entire plants, initially causing foliage bronzing; later branches wilt and sometimes the whole plant eventually dies.

Uncertainties about spider mite reproductive isolation and host range have hindered the wide use of these plant pests in weed control. So, special care must be exercised when these are introduced for weed control. The growing realization that at least some tetranychdis may in fact be quite host specific should promote their future use in weed control.

Table 16.1: Occurrence of Mites on Weeds (Dhooria, 2003)

Mite Species	*Weeds*
Aponychus sulcatus	Narrah
Bryobia neopraetiosa	Bhakoo
Eotetranychus sp.	Calotropis
Eutetranychus orientalis	Cannabis, kaon-makki, basuti, chibbarb, tandla, kundru
Petrobia latens	Mama, *Adhatoda vesica,* button weed
Polyphagotarsonemus latus	Chibbarh, *Bidens biternata, B. pilasa,* amarenthus, salara, makoh, kasundi, sun-berry, karund, asgandh
Tetranychus sp.	Cannabis, amarunthus, makoli.
T. cinnabarinus	Gutpatni, asgandh, puthkanda
T. harti	Mama
T. neocaledonicus	Mama, itsit, wild oat, Madhna
T. urticae	Tandla, karund, billibuti, basuti, pigweed, salara

Eriophyidae

Eriophids are minute mites with elongate, annulate, worm like bodies. These are known as gall, bud, blister or rust mites, depending on their locality and the specific damage they cause to the host plant. This factor was used by some to control unwanted weeds. *Aceria convolvuli*, an eriophid mite was used successfully to control weed, *Convolvulus arvensis* (Rosenthal, 1983). Since this mite readily disperses with air currents special cage was designed for its introduction which prevented its escape. As these mites hibernate in flower heads of weed plant, Kovalev (1973) suggested that these mites can be transported during hibernation.

Tarsonemidae

These are small mites with broad to elongated oval bodies. The integument is hard and shiny. Their diets are highly variable. Some feed on green plants; others prefer fungi and several are associates of arthropods. *Polyphagotarsonemus latus* was found to be associated with weeds growing along with agricultural crops. Certain other members of this family were also recorded from weeds and could, following suitable screening, be considered for the control of these plants.

Galumnidae

These are characterized by large, movable pteromorphs, which are anteriorly rounded and posteriorly pointed. The lamellae are weak or lacking and the cuticle is smooth with small porose areas. *Orthogalumna terebrantis* Wallwork, commonly known as water hyacinth mite feeds on water hyacinth, a floating weed of waterways and lakes. Females lay eggs in the weeds false leaves and immature stages feeds and excavate numerous tunnels in these leaves (Del Fosse, 1978). A female lays 21-24 eggs/week and total development duration from egg to adult is calculated as 10 days (Perkins, 1973). Mite infested weeds becomes prone to pathogenic fungi and phytophagous weevil, *Neochetina eichorniae* (Del Fosse, 1977). The combine activity of both the mite and weevil reduce the water hyacinth density by 50 per cent with in one year (Del Fosse, 1978).

Glossary

Acari: Aristotle used this term for mites, meaning little waxies.

Acarina: Comes under class Arachnida phylum Arthropoda which includes mites and ticks.

Acarology: Study of mites and ticks.

Acarus: Means 'not cut into'. A name given by Linnaeus for 31 mite species in 'Systema Naturae' but now it is a genus of a stored product mite.

Actinotrichida: Group of mites possessing special setae which have internal layer or solid core of optically active material called actinochitin. Astigmata, prostigmata and cryptostigmata come under this category.

Adenotax: The number and distributor of dermal gland pores on the idiosoma.

Adscutal Seta. Seta situated lateral to the scutum.

Aedeagus: The copulatory Sclerotized organ of a male.

Aestivation: A period of inactivity occurring in warm seasons.

Alimentary canal: The digestive tract which consists of fore gut, midgut and hindgut.

Ambulacrum: It consists of pretarsus (post tarsus), pulvillus and claws on leg region.

Anactinotrichida: Group of mites, which lacks optically active material, actinochitin in body setae *e.g.* Mesostigmata, Metastigmata, Notostigmata, Tetrastigmata.

Anal Shield: Shield surrounding the anus.

Anamorphosis: Addition of few segments during the nymphal phase of development.

Anemohoria: Dispersal of mites through air currents.

Anus: The anal opening through which metabolic waste is excreted out. In mites, it is generally ventral and terminal.

Arrhenotoky: Production of males from unfertilized eggs.

Astigmata: Spiracles (stigmata) are absent. Respiration is through cuticle.

Ballooning: A process for dispersal. In some tetranychid species, adults hand themselves through a fine silken thread which acts as parachute and aids in their dispersal.

Brain: A fused circum-oesophageal genglionic mass, also known as synganglion. It is located behind gnathosoma, in the propodosomatic region of idiosoma.

Camerostome: A cavity in the antero-ventral region of propodosoma in which mouth parts can be retracted.

Capitulum: False head. Gnathosoma is some times referred to as capitulum (*e.g.* ticks) because brain is located not in the gnathosomal region but posterior to it.

Caulogastric forms: The tagmata (prosoma and opisthoroma) are connected by a narrow pedicel.

Chaetotoxy: The structure, components, pattern and distribution of setae on body which plays a vital role in mite identification and classification.

Chelicerae: A pair of appendages, in front of the mouth, modified for feeding, usually bears three segments.

Coxa: The basal segment of a leg where it attaches to idiosoma.

Crawling: Means of dispersal with in a host plant/closed storage system or between a concentration of host plants.

Cryptostigmata: Basically soil mites. Highly chitinized beetle like body, stigmata not visible.

Deuterotoky: A form of parthenogenesis in which unfertilized eggs may produce males and females by agamic reproduction.

Deutogyne: All overwintering female population in certain members of Eriophyidae which alternates with protogyne population.

Deutonymph: The second immature eight-legged stage in the life-cycle.

Diapause: It is a neurohormonally mediated state of low metabolic activity which can manifest as delayed morphogenesis, increased resistance to environmental extremes and altered or reduced behavioural activity which begins long before the commencement of unfavourable conditions and terminates well after the ending of such conditions.

Dorsal Shield: Shield partly or completely covering the dorsal surface of the body.

Empodium: Medium claw of the legs is modified into pad like structure.

Epimera: Folds of skin on the ventral side of legs.

Epimorphic: The larva emerging from the egg possesses the definite number of segments.

Epistome: The anterior dorsal prominence of a hood which shields the intersegmented connection between gnathosoma and idiosoma.

Eriophyoidae: Commonly called as bud or gall mites having two pairs of legs in all stages of development.

Euryxenous species: These mite species which use a single host species for dispersal.

Fecundity: The average number of viable eggs laid by a female.

Foregut: Part of alimentary canal which includes pharynx and oesophagus.

Gnathosoma: Body segments bearing mouth and its appendages usually possess three segments, prechericeral, cheliceral and pedipalpal somite.

Grand jean's organ: The anterior lateral edge of the propodosoma is produced forward as a thin cuticular plate which embraces the base of the gnathosoma.

Haemocoel: A blood filled cavity surrounding most of the organs in mites and other orthropods.

Haemolymph: The colourless fluid (blood) of mites and insects filling all the body cavities.

Hexapod stage: A mite stage with three pairs of legs.

Hindgut: Part of elementary canal which is differentiated into posterior intestine, colon, rectal sac and rectum.

Homomorphic male: Male resembles the non gravid female in body shape and length of dorsal setae.

Hypopus: A non-feeding, eight-legged stage developed between the proto–and deuto-nymphal stages in some acarids. It is the dispersal phase in the life cycle of free living Astigmata in which there is complete cessasation of feeding and ontogenesis.

Hypostome: It comprises the endites of pedipalpal coxae.

Hysterosoma: Metapodosoma and opisthosoma region together form hysterosoma.

Idiosoma: A distinct body region which is sac like enclosing the alimentary canal, reproductive organs and nervous system and bears four pair of legs.

Epimera. Chitinised ridges for attachment of muscles, usually situated ventrally.

Integument: A protective exoskeleton comprises of cellular epidermis and non-cellular cuticle.

Labrum: The mouth lying between and below the chelicerae is bordened dorsally by labrum.

Larva: First active stage in the life-cycle and having only six legs.

Latigastric forms: The tagmata (prosoma and opisthosoma) are broadly joined together.

LC_{50}: It is an amount of poison per unit weight of the organisms expressed as µg/ g of body weight. It is the lethal dose of toxicant which kills 50 per cent of test population.

LT_{50}: The time required to kill 50 per cent of test population at a certain dose or concentration.

Malphigian tubules: The chief excretory organ which arise from post ventricular region or at the function of mid and hindgut in mites.

Mesostigmata: Pair of stigmata is situated between coxae III and IV of legs on idiosoma.

Metapodosoma: Body segments bearing third and fourth pairs of legs.

Metastigmata: Presence of pair of stigmata behind coxae IV of legs on idiosoma.

Microsolonidia: Setae usually confined to tarsi of first pair of legs.

Midgut: Part of alimentary canal which includes stomach, entestine. It is the area of absorption.

Multivoltine species: More than one generation are completed with in a year. Generally occur in domestic, plant and certain soil species.

Neosomy: A radical intrastadial metamorphosis which results in new body structures as a consequence of extensive cuticular growth.

Octapod stage: A mite stage with four pairs of legs.

Opisthosoma: Body segments posterior to fourth pair of legs.

Oviposition: A specialized egg laying organ in females, formed from outgrowths of idiosomal segments.

Parthenogenesis: Development of an unfertilized ovum into a new individual genetically identical with the parent.

Pedipalp: A pair of sensory appendages situated behind the mouth. It is a second pair of appendages present on gnathosoma bearing five segments which are named in the same sequence as those of true legs. Modified in few families.

Peritreme: Long, anteriorly directed grooves associated with stigmata (spiracles).

Phoresy: A phenomenon in which one animal actively seeks out and attaches to the outer surface of another animal for. a limited time for transportation and dispersal.

Pleomorphic male: Male has larger body, longer dorsal setae and modified third pair of legs.

Podomeres: The leg segments of a mite or an insect. In adults six segments *viz.* coxa, trochanter, femur, genu, tibia and tarsus are usually present.

Podosoma: Leg bearing region of idiosoma.

Poroidotaxy: The number and distribution of stress sensitive organules of poroids on the idiosoma.

Prelarva (Deutovum): It is an inactive, preecolosion stage which exists with in the egg shell and characterized by its distinct embryonic cuticle.

Pretarsus (=Post tarsus): The last segment of the tarsus, usually bears claw.

Propodosoma: Body segments bearing first and second pairs of legs.

Prostigmata: Pairs og stigmata are located near gnathosoma onidiosoma.

Proterosoma: Gnathosoma and propodosoma region together form proterosoma.

Protogyne: Presence of both the sexes in the progeny in eriophyid population which alternates with deutogynes.

Protonymph: First, immature, eight-legged stage in the life cycle.

Pseudostigmata: Modified branched setae or balloon or club shaped present at the anterior dorsal body. Present in cryptostigmata.

Pteromorphae: Immovable or movable wing like projections in Oribatida which in well developed state bend ventrally to protect the posterior legs.

Pulvillus: Membranous structure associated with the claws of the legs.

Quiescence: It is a state of low metabolic activity which occurs in direct response to adverse physical conditions and is terminated as soon as favourable condition returns.

Resurgence: Numerical increase in pest populations after acaricide/ insecticide treatment, for above those occurring in conventionally treated plots.

Scabellum: A plateform like structure below the vertex in some species (Uropodina) to accommodate first pair of legs.

Scapular Seta: Seta situated dorsally in the posterior region of the propodosoma in some acarids.

Schizeckenosy: A process in which hind gut in absence of anus, expel accumulated faecal material/metabolic waste by means of direct rupture through exoskeleton.

Scutum: Chitinised shield-like structure dorsally near the anterior end of the body.

Silk: A tough material in the form of extremely tine threads secreted by silk glands in tetranychid species.

Solenidion: Chemosemory setae. It is a thin chitinous tube without a swollen base open at its proximal end and present on distal podosomal segments.

Spanandric male: Production of non functional males by thelytokons mites.

Stermites: A single shield or carapace showing varying degrees of fusion on the ventral surface of prosoma.

Sternal Shield: Chitinised plate between the basal segments of the second pair of legs in the Mesostigmata.

Sternal-Genital Shield: A compound shield occupying the region between the legs in the males of some Mesostigmata.

Stigmata: The spiracular or respiratory opening whose relative position in mite body is a major diagnostic feature for the separation of suborders with in Acarina.

Sucker Plate: A plate provided with suckers situated postero ventrally in the hypopus.

Tagma (Plural Tagmata): A group of segments farming a unit in the body of an arthopod. In mites, there are two tagmata *viz.* gnathosoma (Jaw region) and idiosoma (body region having locomotory appendages).

Tarsonemidae: Commonly called as broad and yellow phytophagus mites which are fast moving.

Tenuiplilpidae: Commonly called as false spider mites which look 1ike spider mites but cannot spin webs.

Tergites: A single shield or carapace showing varying degrees of fusion on the dorsal surface of prosoma.

Tetranychidae: Commonly called as spider mites as they can spin webs to protect their colonies from natural enemies and pesticides.

Thanatosis: When attacked or disturbed, mites draw their appendages inside and remain motionless. When damage is over, they resume their normal activity *e.g.* Trombidi mites.

Thelytoky: Production of females from unfertilized eggs.

Toxicant: It is an agent capable of interfering adversely with vital process of organisms by physiological means.

Trichobothria: Setae modified as special tactile hairs, present on idiosoma and legs.

Tuckerellidae: Brighlty coloured phytophagus mites having fan like dorsal body setae and long whip like caudal setae.

Univoltine Species: One generation per year is produced by a species. Generally occur in soil, litter and some aquatic habitats.

Ventri-Anal Shield: A compound shield situated posterior to the fourth pair of legs in the Mesostigmata.

Zoohoria: Utilization of other animals for dispersal/transportation.

References

Accorti, M., Borbattini, R. and Marchetti, S. 1986. La diagnosi ed il controllo di *Varroa jacobsoni* Dud. in campo: proposta di unificazione delle metadologie melle prove sperimentali. *Apieoltura* 2: 165-185.

Addison, J.A., Hardman, J.M. and Walde, S.J. 2000. Pollen availability for predaceous mites on apple: spatial and temporal heterogencity. *Exp. App. Acarol.* 24: 1-18.

Aggarwal, K. and Kapil, R.P. 1988. Observations on the effect of queen cell construction on *Euvarroa sinhai* infestation in drone brood of *Apis florea*. In: *Africanized Honey Bees and Bee Mites*. (eds. Needham, G.R., Page, R.E., Delfinado-Baker, M., Bowman, C.E.). Ellis Horwood Pub., U.K.: 404-408.

Aggarwal, K. and Kapil, R.P. 1989. Seasonal population dynamics of *Tropilaelaps clareae* (Acari: Laelapidae) in *Apis dorsata* colonies. In: *Progress in Acarology* (eds Channabassavanna, G.P. and Viraktamath, C.A.), Oxford and IBH Publ. Co. Pvt. New Delhi. 2: 283-286.

Al-Amidi, A.H.K.and Downes, M.J. 1990. *Parasitus hituherosus* (Acari: Parasitidae), a possible agent for biological control of *Heteropeza pygmaea* (Diptera: Cecidomyiidae) in mushroom compost. *Exp. Appl. Acarol*. 8: 13-25.

Al-Amidi, A.H.K, Dunne, R. and Downes, M.J. 1991. *Parasitus bituberosus* (Acari: Parasitadae): An agent for control of *Lycoriella solani* (Diptera. Sciaridae) in mushroom crops. *Exp. Appl. Acarol.* 11: 159-166.

Al-Badry, E.A., Rizk, G.N., Hafez, S.M. and Al-Badry, E.A. 1980. Biological studies on the predatory mite, *Acaropsis docta* (Berlese) attacking stored product insects. *Mesopotamia J. Agric.* 15 (1): 179-202.

Alenear, J.A., Haji, F.N.P., Moreira, F.R.B. and Alenear, J.A. 1999. Coconut mite *Aceria guerreronis* (Keifer) biological aspects, symptoms, damage and control methods. *Indian Coconut J.* 30: 18.

Allen, M.D. 1965. The effect of a plentiful supply of drone comb on colonies of honeybees. *J. Apic. Res.*: 109-119.

Alwar, V.S. 1970. Feather mites of poultry in India. *Ceylon Vet. J.* 18 (3): 61-64.

Amano, H. and Chant, D.A. 1977. Life history and reproduction of two species of predacious mites, *Phytoseiulus persimilis* Athias-Henriot and *Amblyseius andersoni* (Chant) (Acarina: Phtoseiidae). *Can. J. Zool.* 55: 1978-1983.

Amano, H. and Chant, D.A. 1978. Some factors affecting reproduction and sex ratios in two species of predecious mites, *Phytoseiulus persimilis* Athias Hemiot and *Amblyseius andersoni* (Chant) (Acarina: Phytoseiidae). *Can. J. Zool.* 56: 1593-1607.

Amin, A.W., Mowafe, M.H. and Fatma, S.A. 1999. Effect of predaceous mesostigmatid mites in the control of *Meloidogyne javanica*, root–knot nematode on kidney bean. *Pakistan J. Nematol.* 17: 91-96.

Ancons, G. 1923. Asma epidemico da "Pediculoides velllricosus" *Policlnico* 30: 45-70.

Anderson, A. 1988. Population growth and developmental stages of the house dust mite, *Dermatophagoides pteronyssinus* (Acari: Pyroglyphidae). *J. Med. Entomol.* 25: 370-373.

Anderson, J.R. 1983. Mites as biological control agents of dung breeding pests: practical considerations and selection for pesticide resistance. In: *Biological Control of Pests by Mites* (eds. Hoy, M.A., Cunningham, G.L. and Knutson, L.) Univ. California: 99-102.

Anderson, R.C. 1968. Ecological observations on three species of *Pimeliaphilus* parasites of Triatominae in the United States. (Acarina: Pterygosomidae) (Hemiptera: Reduviidae). *J. Med. Entomol.* 5: 459-464.

Andres, L.A. 1957. *An ecological study of three species of Tetranychus (Acarina: Tetranychidae) and their response to temperature and humidity.* Ph.D dissertation, University of California, Btrkeley.

Andres L.A. 1983. Considerations in the use of phytophagous mites for biological control of weeds. In: *Biological Control of Pests by Mites* (eds. Hoy, M.A., Cunningham, G.L. and Knutson, L.): 53-56.

Anil, K.N., Mallik, B.S. and Kumar, H.M. 1995. *Abst IV Nat Symp. Acarology*. Bangalore, P. 72-73.

Anonymous. 1984. Recommended methods for the detection and measurement of resistance of agricultural pests to pesticides. Methods for Phytoseiid predatory mites–FAO method No. 32. *FAO Plant Protect. Bull.* 32: 25-27.

Archer, T.L., Murray, C.L., Eikenbary, R.D., Starks, K.J. and Morrison, R.D. 1973. Cold storage of *Lysiphlebus testaceipes* mummies. *Environ. Entomol.* 2: 1104-1108.

Arlian, L.G. 1989. Biology and ecology of house dust mites, *Dermatophagoides* spp. and *Euroglyphus* spp. *Immunol. Allergy Clinics N. Am.* 1: 339-350.

Arlian, L.G. 1991. House dust mite allergens: a review. *Exp. App. Acarol.* 10: 167-186.

Arlian, L. G. 1992. Water balance and humidity requirements of house dust mites. *Exp. Appl. Acarol.* 16: 15-35.

Arlian, L.G. and Veselica, M.M. 1979. Water balance in insects and mites. *Comp. Biochem. Physiol.* 64: 191-200.

Arlian, L.G. and Wharton, G.W. 1974. Kinetics of active and passive components of water exchange between the air and a mite, *Dermatophagoides farinae. J. Insect Physiol.* 20: 1063-1077.

Arlian, L.G, Rapp, C.M. and Ahmed, S.G. 1990. Development of *Dermatophagoides pteronyssinus* (Acari: Pyroglyphidae). *J. Med. Entomol.* 27: 1035-1040.

Arlian, L.G., Vyrzenski-Mother, D.L. and Pole, M.J. 1989. Survival of adults and developmental stages of *Sarcoptes scabiei* Val'. Canis when off the Host. *Exp. Appl. Acarol.* 6: 181-187.

Arlian, L.G., Woodford, P.I., Benistein, I.L. and Gallaghen, J.S. 1983. Seasonal population structure of house dust mites, *Dermatophagoides* sp. (Acari: Pyroglyphidae). *J. Med. Entomol.* 20: 99-102.

Arthur, F.S. and Axtell, R.C. 1983. Susceptibility of northern foul mites in North Carolina to five acaricides. *Poult. Sci.* 62: 428-432.

Asquith D. 1965. Photography in mite counting. *J. Econ. Entomol.* 58: 677-678.

Atwal, A.S. 1967. Studies on the acarine disease problems in *Apis indica. XXI Int. Beekeeping Cong. Prelim. Sci. Meet.* Summary Paper 45: 70-71.

Atwal, A.S. 1969. Serious infestations of *Apis mellifera* L. bee by the ectoparasitic mite *Tropilaelaps c1areae. XXII lnt. Beekeeping Cong.*, Summary 22: 108.

Atwal, A.S. 1971. Acarine disease problem of the Indian honey bee, *Apis indica F. Am. Bee J.* III: 134-135,186-187.

Atwal, A.S. and Goyal, N.P. 1971. Infestation of honey bee colonies with *Tropilaelaps* and its control. *J. Apic. Res.* 10: 137-142.

Atyeo, W.I. and Perez, I.M. 1988. Species in the genus *Rhytidelasma gaud* (Acarina: Pterolichidae) from the green conure, *Aratinga Lolochlora* (Sclater) (Aves: Psittacidae). *System. Parasitol.* 11: 85-96.

Atyeo, W.I. and Perez, T.M. 1990. Feather mites of the *Aralichus canestrinii* (Trouessasrt) complex (Acarina, Pterdichidae) from New World Parrots (Psittacidae) II From the genera *Aratinga Spix, Deroptyus wasgler, Letposittaca Bertepsch* and *Stolzmann, Ognorhynchus Bonaparte, Pionites heine* and *Pyrrhura Bonaparte,* and conclusions to the study. *Fieldiana Zool.* 62: 1-30.

Avella, M., Fournier, D., Pralavorio, M. and Berge, J.B. 1985. Selection pour la resistance a la deltamethrin d'ure sonche de *Phytoseiulus persimilis* Athias-Henriot. *Agronomic* 5: 177-180.

Axtell, R.C. 1966. Comparative toxicities of insecticides to house fly larvae and *Macrocheles muscaedomesticae*, a mite predator of the house fly. *J. Econ. Entomol.* 59: 1128-1130.

Axtell, R.C. and Arends, J.J. 1990. Ecology and management of arthropod pests of poultry. *Ann. Rev. Entomol.* 135: 101-126.

Axtell, R.C. and Rutz, D.A. 1986. Role of parasites and predators as biological fly control agents in poultry production facilities. *Misc. Publ. Entomol. Soc. Am.* 61:88-100.

Baker, E.W. and Wharton, G.W. 1952. *An Introduction to Acarology*. The MacMillan Company, New York. 465 pp.

Baker, G.T. 1996. Chemoruption in four species of water mites, (Acari: Hydrachnida): behavioural and morphological evidence *Exp. Appl. Acarol.* 20: 445-455.

Baker, J.T. and Nemizig, H.N. 1968. *Hirsutella thompsonii* as a fungus parasite of the blue berry loud mite. *J. Econ. Entomol.* 61: 1117-1118.

Banu, K. and Channabasavanna, G.P. 1972. Plant feeding mites of India: Preliminary account of the biology of the spider mite, *Eutetranychus orientalis* (Klein) (Acerina: Tetranychidae). *Mysore J. Agric. Sci.* 6: 253-268.

Baradaran, P., Arbabi, M., Kamali, K. and Ostovan, H. 2001. Study on the biology of *Tetranychus cinnabarinus* (Boisd.) on ornamental plants under laboratory conditions. *Appl. Ent. Phytopath.* 69: 9-10.

Bartsch, I. 1996. Halacaroids (Halacaroidea, Acari) in freshwater. Multiple invasions from the Paleozoic onwards *J. Nat. Hist.* 30: 67-99.

Beavers, J.B. and Reed, D.K. 1972. Susceptibility of seven tetranychids to non-occluded virus of the citrus real mite and the correlation of the carmine spider mite as a vector. *J. Invert. Pathol.* 20: 279-283.

Beckendorf, S.K. and Hoy, M.A. 1985. Genetic improvement of arthopod natural enemies through selection hybridization or genetic engineering techniques. In: *Biologial Control in Agricultural IPM, Systems* (eds. Hoy, M.A., Herzog, D.C.), Academic Press, New York: 167-187.

Behnke, J.M., McGregor, P.K., Shepherd, M., Wiles, R., Barnard, C., Gilbert, F.S. and Hurst, J.L. 1995. Identity, prevalence and intensity of infestation with wing feather mites on birds (Passeriformes) from the Setubal Peninsula of Portugal. *Exp. App. Acarol.* 19: 443-458.

Berlese, A.1905. Apparecchio per raccoglieve presto ed in gran numero piccoli artropodi. *Redia* 2: 85-89.

Berreen, J.M. 1984. The functional response of *Cheyletus eruditus* Schrank to changes in the density of its prey, *Acarus siro* L. In: *Acarology VI* (eds. Griffiths, D.A. and Bowman, CE.), Ellis Horwood, Chichester, 2: 980-986.

Berry R.E. 1973. Biology of the predaceous mite, *Pergamasus quisquiliarum* on the garden symphylan, *Scutigerella immaculata* in the laboratory. *Ann. Entomol. Soc. Am.* 66: 1354-1356.

Bhanot, J.P.2003. Biology and feeding behaviour of predatory mite, *Acaropsis docta* (Berlese) (Acarina: Cheyletidae). In: *Mites, their identification and management* (eds. Yadav, P.R., Chauhan R., Putatunda, B.N. and Chhillar, B.S.), Dept. of Entomology, CCSHAU, Hisar: 1-3.

Bharadwaj, R.K. 1968. A new record of mite, *Tropilaelaps clareae* from *Apis dorsata*. *Bee Wld.* 49: 115.

Bhaskar, S. and Putatunda, B.N. 1989. Mesostigmatid mites associated with bees in India. In:*Progress in Acarology* (eds. Channabasavanna, G.P. and Viraktamath, C.A.), Oxford and IBH Publ. Co. Pvt. Ltd., New Delhi 2: 287-289.

Bilgrami, R.S. 1982. Important fungal diseases of Litchi and their control. In: *Special issue on litchi*, GADA, Muzaffarpur, Pub. No. 14/81: 50-57.

Bilgrami, A.L. 1994. Predatory behavior of a nematode feeding mite *Tyrophagus putrescentiae* (Sarcoptiformes: Acaridae). *Fundam Appl. Nematol.* 17: 293-297.

Bilgrami, A.L. and Tahseen, Q. 1992. A nematode feeding mite *Tyrophagus putrescentiae* (Sarcoptiformes: Acaridae). *Fundam. Appl. Nematol.* 15: 477-478.

Binns, E.S. 1973. Predatory mites-neglected allied? *Mushroom J.* 12: 5640-5644.

Bischoff, E. 1988. Methodes actuelles de quantification des acariens dans 1 habitat. *Rev. Franc. Allergol.* 28: 115-122.

Bischoff, E. and Von Bronswijk, J.E.M.H. 1986. Beitrage zur Okologie derttausstaubmilben.1. Uber die Frreichbarkeit von Hausstaubmilben durch Absangen. *Allelgologie* 9: 375-378.

Bischoff, E.R.C., Fischer, A. and Liebenberg, B. 1992. Assessment of mite numbers: new methods and results. *Exp. Appl. Acarol.* 16: 1-14.

Blainey, A.D., Topping, M.D., Ollier, S. and Davies, R.J. 1989. Allergic respiratory disease in grain workers: The role of storage mites. *J. Allergy Clin. Immunol.* 84: 296-303.

Blythe, M.E. 1976. Some aspects of the ecological study of housedust mites. *Br. J. Dis. Chest.* 70: 3-31.

Blythe, M.E., Williams, J.D. and Morrison, S. J. 1974. Distribution of pyroglyphid mites in Birmingham with particular reference to *Euroglyphus maynei*. *Clin. Allergy.* 4: 25-33.

Boczek, J. 1975. Reproductive biology of *Tyrophagus putrescentiae* (Schrunk) (Acarina: Acaridae). In: *Proc. 1st Instl. Waking Conf. Stored Prod. Ent.*, Sevannah, USA: 154-159.

Boczek, J. and Czajkowska, B. 1968. Main sources of infestation of stored grain by mites. *Priz. Zboz-mlyn.* 12(9): 275-279.

Boczek, J. and Czajkowska, B. 1973. A study on some aspects of ageing in *Acarus siro* L. (Acarina: Acaridae). *Ekol. Pol.* 21(11): 173-184.

Boot, W.J., Colis, J.N.M. and Beetsma, J. 1992. Differential periods of *Varroa* mite invasion into worker and drone cells of honeybees. *Exp. App. Acarol.* 16: 295-301.

Borah, D.C. and Rai, P.S. 1989. In: *Progress in Acarology* (eds. Channabasavanna, G.P. and Viraktamath, C.A.), Oxford and IBH Publ. Co. Pvt. Ltd., New Delhi 2: 375: 379.

Borneck, R. 1985. Sur le front de la Varroatose. *Rev. Francaise d' apiculture* 458: 556-567.

Bot, J. and Meyer, M. 1969. An artificial rearing medium for acarid mites. *J. Ent. Soc. Sth. Afr.* 29: 199.

Bower, C.C. 1977. European red mite-apple pest in N. S. W. *Agric. Gaz. NSW* 88 (5): 40-41.

Bowie, M.H., Worner, S.D. and Chapman, R.B. 1999. The use of image analysis to study the effects of residues of fenvalerate on the locomotry behaviour of *Panonychus ulmi* and *Typhlodromus pyri* (Acari: Tetranychidae, Phytoseiidae). *Exp. App. Acarol.* 23: 1-9.

Boyd, E..M. 1951. The external parasites of birds: A review. *Wilson Bull.*, 63 (4): 363-369.

Braude, R.A.G., Low, A.G., Mitchell, K.G., Pittman, R.J. and Wilkin, D.R. 1980. Effect of flour mite infestation (*Acarus siro* L.) on nutritive value of pig diets. *Vel. Rec.* 106: 35-36.

Braun, A.R., Guerrero, J.M., Bellotti, A.C. and Wilson, L.T. 1989. Within plant distribution of *Mononychellus tanajoa* (Bondav) (Acari. Tetranychidae) on cassava: effect of clone and predation on aggregation. *Bull. Ent. Res.* 79: 235-249.

Bruce, W.A. 1983. Mites as biological control agents of stored product pests. In: *Biological Control of Pests by mites* (eds. Hoy, M.A., Cunningham, J.L. and Knutson, L.V.). Div. of Agric. Nat. Res, Univ. of California, Beskeley, Spec. Publ. 3304: 74-78.

Bruce, W.A. 1989. Artificial Diet for the parasitic mite *Pyemotes tritici* (Acari. Pyemotidae). *Exp. Appl. Acarol.* 6: 11-18.

Bruce, W.A. and Le Cato, J. L. 1979. *Pyemotes tritici*: Potential biological control agent of stored product insects. In: *Recent Advances in Acarology* (ed. Rodriguez, J.G.). Academic, New York, 1: 213-220.

Bruce, W.A. and LeCato, J.L. 1980. *Pyemotes tritici*: A potential new agent for biological control of the red imported fire ant, *Solellopsis invicta* (Acari: pyemotidae). *Int. J. Acarol.* 6 (4): 271-274.

Bruce-Olivier, S.J. and Hoy, M.A. 1990. Effect of prey stage on life table attributes of a genetically manipulated strain of *Metaseiulus occidentalis* (Acari: Phytoseiidae). *Exp. Appl. Acarol.* 9: 201-217.

Brust, G.E. and House, G..J. 1988. A study of *Tyrophagus putrescentiae* (Acari: Acaridae) as a facultative predator of southern corn root worm eggs. *Exp. Appl. Acarol.* 4: 335-344.

Burgett, D.M., Rossignol, P.A. and Kitprasert, C. 1990. A model of dispersion and regulation of brood mite *(Tropilaelaps clareae)* parasite on the giant honey bee *(Apis dorsata) Canadian J. Zool.* 68: 1423-27

Burgett, M., Akaratanakul, P. and Morse, R.A. 1983. *Tropilaelaps clareae:* a parasite of honey bees in South-east *Asia. Bee Wld* 64(1): 25-28.

Burr, M.L., Neale, E., Dean, B.V. and Verrier-Jones, E.R. 1980. Effect of a change to mite free bedding on children with mite sensitive asthma: a controlled trial. *Thorax*. 35: 513-514.

Calatayud, F. and Verdu, M.J. 1993. Hive debris counts in honeybee colonies: a method to estimate the size of small populations and rate of growth of the mite *Varroa jacobsoni* Oud. (Mesostigmata: Varroidae). *Exp. App. Acarol.* 17 (12): 889-894.

Calatayud, F. and Verdu, M.J. 1994. Survival of the mite *Varroa jacobsoni* Oud. (Mesostigmata. Varroidae) in broodless colonies of the honeybee *Apis mellifera* L. (Hymenoptera: Apidae). *Exp. App. Acarol.* 18 (10): 603-612.

Castagnoli, M., Liguori, M. and Nannelli, R. 1989. A methodology for rearing *Euroglyphus maynei* (Coar.) (Acarina Pyroglyphidae) in laboratory. *Redia*. 72: 127-131.

Cayrol, J. C. 1970. Action des autres composants de la biocenose du champignon de couche sur le nematode mycophage, *Ditylenchus myceliophagus* J.B. Goodey, 1958, et etude de son anabiose: forme de survie en conditions defavourables, *Rev. Ecol. Biol. Sol.* 7: 409-440.

Cerny, V. 1973. Parasite-host relationships in feather mites. In: *Proc. III Intl. Cong. Acarol.* (eds. Daniel, M. and Rosickly, B.). Academic Publishing House, Prague: 761-764.

Channabasavanna, G.P. 1966. *A contribution to the knowledge of Indian eriophyid mites (Eriophyoidea: Trombidiformes, Acarina).* Univ. Agric. Sci. Bangalore: 130.

Channabasavanna G.P. 1981. *Contributions to Acarology in India.* Acarological Society of India, Bangalore. 256 pp.

Channabasavanna G.P. and Viraktamath, C.A. 1988. *Progress in Acarology*. Oxford and IBH Publ. Co. Pvt. Ltd., New Delhi, (Vol I and II): 511, 484 pp.

Chapman, R.B., Penman, D.R. and Walker, J.T.S. 1987. European red mite resistance to organotin miticides in Hawke's Bay apple orchards. *Proc. Z.Z. Weed Pest Control Conf.* 40: 94-98.

Chatterjee, K.S. and Gupta, S.K. 1996. *J. Beng. Nat. Hist. Soc.* 15 (2): 18-27.

Chaudhary, A. 1991. *Studies on mites associated with small mammals in Hisar.* M.Sc. Thesis, CCSHAU, Hisar.

Chen, J.S. 1990. An improved method for determining the susceptibility of *Rhizoglyphus robini* and *R. setosus* (Acarina: Acaridae) to pesticides. *Exp. App. Acarol.* 8: 175-178.

Chen, J.S. and Lo, K.C. 1989. Susceptibility of two bulb mites, *Rhizoglyphus robini* and *R. setosus* (Acarina: Acaridae) to some acaricides and insecticides. *Exp. App. Acarol.* 6: 55-66.

Chen, P.R., Zhang, Z.Q., Wang, K., Wang, X.Y., Xu, W.L. and Gao, Z.L. 1994. *Allothrombium pulvinum* Ewing (Acari: Trombidiidae), an important early season natural enemy of *Aphis gossypii* (Hom., Aphididae) in cotton. *J. Appl. Entomol.* 116: 113-121.

Chiang, H.C. 1970. Effects of manure applications and mite predation on corn rootworm populations in Minnesota. *J. Econ. Entomol.* 63: 934-936.

Choe, J.C. and Kim, K.C. 1987. Community structure of arthropod ectoparasites on Alaskan seabirds. *Can. J. Zool.* 65: 2998-3005.

Choe, J.C. and Kim, K.C. 1988. Microhabitat preference and coexistence in feather mites (Acari-Analgoidea) on Alaskan seabirds. *Oecologia.* 79: 10-14.

Chua, K.Y., Doyle, C.R., Simpson, R.J., Turner, K.J., Stewart, G.A. and Thomas, W.R. 1990. Isolation of cDNA coding for the major mite allergen Der p II by IgE plaque immunoassay. *Int. Arch. Allergy Appl. Immunol.* 91: 118-123.

Cicilykutty, A.K., Balasubramaniam, N.K. and John, P.A. 1981. The influence of physical conditions (salt content) of the flesh of fish on the intensity of infestation by adult mite, *Suidasia medanensis. Acarologia* 22(3): 291-294.

Cicolani, B. 1979. The intrinsic rate of natural increase in dung macrochelid mites, predators of *Musca domestica* eggs. *Boll. Zool.* 46: 171-178.

Clark, E. W. and Morishita, F. 1950. C-M medium: A mounting medium for small insects, mites and other whole mounts. *Science* 112: 789-790.

Clements, D.R. and Hormsen, R. 1993. Prey preferences of adult and immature *Zetzellia mali* Ewing (Acari: Stigmaeidae) and *Typhlodromus candiglans* Schuster (Acari: Phytoseiidae). *Can. Entomol.* 125: 967-969.

Cochran, W.G. 1977. *Sampling techniques* 3rd ed. John Wiley and Sons, New York. 428pp.

Cohen, E. and Gersson, U. 1989. Resurgence of spider mites (Acari: Tetranychidae) induced by Synthetic Pyrethroids. *Exp. Appl. Acarol.* 6: 29-46.

Cohen, M.B. 1928. Further observations on the use of filtered air in the diagnosis and treatment of allergic conditions. *J. Lab. Clin. Med.* 60: 963-967.

Colloff, M.J. 1987. Mites from house dust in Glasgow. *Med. Vet. Entomol.* 1: 163-168.

Colloff, M.J. 1988. Mite ecology and microclimate in my bed. In: *Mite Allergy-A worldwise problem* (eds. DeWeek, A.L. and Todt, A.). *Int. meet. Bad Kreuznach*, Germany, 1-2 September, 1987, UCB, Brussel: 51-54.

Colloff, M.J. 1989. A New and Rapid Method of Making Permanent preparations of large number of house dust mites for light microxcopy. *Exp. Appl. Acarol.* 7: 323-326.

Colloff, M.J. 1991. A review of biology and allergenicity of the house-dust mite *Euroglyphus maynei* (Acari: Pyroglyphidae). *Exp. Appl. Acarol.* 11: 177-198.

Cook, D.A. and Armitage, D.M. 1999. The efficacy of Dryacidae, an inert dust, against two species of Astigmated mites, *Glycyphagus destructur* and *Acarus siro* at nine temperature and moisture content combinations on stored grains. *Exp. Appl. Acarol.* 23(1): 51-63.

Cook, D.R. 1976. Water mite genera and subgenera. *Mem. Amer. Ent. Inst.* 21: 1-860.

Cook, D.R. 1980. Studies on Neotropical water mites. *Mem. Amer. Ent. Inst* 31: 1-645.

Cook, D.R. 1986. Water mites from Australia. *Mem. Amer. Ent. Inst.* 40: 1-568.

Coop, L.B. and Croft, B.A. 1995. *Neoseiulus fallacis*: dispersal and biological control of *Tetranychus urticae* following minimal inoculations into a strawberry field. *Exp. App. Acarol.* 19: 31-43.

Crane, E. 1968. Mites infesting honeybees in Asia. *Bee Wld* 49 (3):113-114.

Croft, B.A. and McMurtry, J.A. 1972. Comparative studies on four strains of *Typhlodromus occidentalis* Nesbitt (Acarina: Phytoseiidae) IV Life history studies. *Acarologia* XIII (3): 460-470.

Croft, B.A. 1972. Prey stage distribution, a factor affecting the numerical response of *Typhlodromus occidentalis* to *Tetranychus medanielli* and *Tetranychus pacificus*. *Great Basin Nat*. 32: 61-75.

Croft, B.A. 1990. *Arthropod biological control agents and pesticides*. John Wiley and Sons, New York.

Cromroy, H.L. 1983. Potential use of mites in biological control of terrestrial and aquatic weeds. In: *Biological control of pests by mites* (eds Hoy, M.A., Cunningham, G.L. and Knutson, L.): 61-66.

Crystal, M.M. and DeMilo, A.B. 1984. A laboratory test method for evaluation of acaricides against northern fowl mites. *J. Econ. Entomol.* 57: 602-604.

Cunnington, A.M. 1965. Physical limits for the complete development of the grain mite *Acarus siro* L. (Acaridae: Acarina) in relation to its world distribution. *J. Appl. Ecol.* 2: 295-306.

Cunnington, A.M. 1967. The mite fauna of house dust (Proc. Br. Allergy Soc.) *Acta Allerg.* (kbh). 22: 415.

Currie, G.A. 1934. The bdellid mite *Biscirus lapidaries* Kramer, predatory on the Lucerne flea *Sminthurus viridus* L. in western Australia. *J. Aust. Counc. Sci. Ind. Res.* 7: 9-20.

Cuthbert, O.D., Brostoff, J., Wraith, D.G. and Brighton, W.D. 1979. "Barn allergy" asthma and rhinitis due to storage mites. *Clin. Allergy*. 9: 229-236.

D'Aguilar, J., Benard, R. and Bessard, A. 1957. Une method de lavage pour extraction des arthropods terricoles. *Ann. Epiphyt.* 8: 91-99.

Daniel, M.1981. In: *Contributions to Acarology in India*: 167-173.

Danka, R.G., Villa, J.D., Rinderer, T.E. and De Latte, F.T. 1995. Field test of resistance to *Acarapis woodi* (Acari: Tarsonemidae) and of colony production by four stocks of honey bees (Hymenoptera: Apidae). *J. Econ. Entomol.* 88: 584-91.

Dar, N. and Gupta, V. 1979. Studies on the house dust mites of India and their role in causation of Bronchial Asthma and allergic rhinitis. *Oriental Insects.* 13: 261-298.

Davids, C. and Schoots, C. 1975. The influence of the water mite species *Hydrachna conjecta* and *H. cruenta* (Acari, Hydrachnellae) on the egg production of the Corixidae *Sigara striata* and *Cymatia coleoptrata* (Hemiptera). *Verh. Int. Verein. Limnol.* 19: 3079-3082.

Davidson, N.A., Dibble, J.E., Flint, M.L., Marer, P.J. and Guye, A. 1991. *Managing insects and mites with spray oils.* University of California, Oakland.

DeAngelis, J.D., Larson, K.C., Berry, R.E. and Krantz, G.W. 1982. Effects of spider mite injury on transpiration and leaf water status in peppermint. *Environ. Entomol.* 11: 975-978.

DeJong, D., Morse, R.A. and Eickwort, G.C. 1982. Mite pests on honeybees. *Ann. Rev. Entomol.* 27: 229-252.

DeJong, D. 1988. *Varroa jacobsoni* does reproduce in worker cells of *Apis cerana* in South Korea. *Apidologie* 19: 241-244.

Del Cerro, M., Cogen, J. and Del Cerro, C.1980. Stevenel's blue, an excellent stain for optical microscopical study of plastic embedded tissues. *Microsc. Acta* 83: 117-121.

Del Fosse, E.5. 1977. Effect of Orthogalumna terebrantis (Acari: Galumnidae) on *Neochetina eichhornine* (Col. Curculionidae) eggs and oviposition. *Entomophaga* 22: 359-63.

Del Fosse, E.S. 1978. Interaction between the mottled waterhyacinth weevil, *Neochetina eichhorniae* Warner and the waterhyacinth mite, *Orthogalumna terebrantis* Wallwork. In: *Proc. Int. Symp. Bioi. Cant. Weeds,* (ed. Freeman, T.E.), University of Florida: 93-97.

Delfinado, M.D. and Baker, E W. 1974. Varroaidae, a new family of mites in honey bees. J. *Washington Acad. Sci.* 64 (1):4-10.

Delfinado-Baker, M. and Baker, E.W. 1982. A new species of *Tropilaelaps* parasitic on honeybees. *Am. Bee J.* 122 (6):416-417.

Delfinado-Baker, M., Underwood, B.A. and Baker, E.W. 1985. The occurrence of *Tropilaelaps* mites in brood nests of *Apis dorsata* and *A. laboriosa* in Nepal, with descriptions of the nymphal stages: *Am. Bee J.* 125:703-706.

DeshRaj, Sharma, S.D., Sharma, O.P. and Singh, M. 1991. Apparent symptoms oftracheal mite infestation in the colonies of *Apis mellifera* L. *Indian Bee J.* 53 (1-4):78-79.

Dhaliwal, H.S., Srivastava, S. and Sharma, P.L. 1974. Some biological observations on the mite, *Acarapis woodi* Rennie infesting honey bee, *Apis cerana indica* F. *Curr. Sci.* 43 (23):750-752.

Dhooria, M.S. 1980. *Acar. Newsl.* 10: 4-6.

Dhooria, M.S. 2003. Mite pests and their management in Punjab. In: *Mites, their identification and management* (eds. Yadav, P.R., Chauhan, R., Putatunda, B.N. and Chhillar, B.S.) Centre of Advanced Studies, Department of Entomology, CCSHAU, Hisar (India): 41-47.

Dhooria, M.S., Bhullar, M.B. and Mallik, B. 2005. *Mite pests of citrus and their management in India.* AINP (Agricultural Acrology), UAS, Banglore: 28pp.

Dhooria, M.S., Kapoor, S.K. and Kapur, S.P. 1987. New record of tarsonemid mite, *Polyphagotarsonemus latus* (Banks) (Acari: Tarsonemidae) as a pest of citrus in India. In: *Contribution to Acarological Researches in India* (eds. Mukherjee, A.B. *et al.*), Dept. Agril. Entomology, BCKV, W. Bengal: 325-328.

Dicke, M., Sabelis, M.W. and DeJong, M. 1988. Analysis of prey preference in phytoseiid mites by using an olfactometer, predation models and electrophoresis. *Exp. Appl. Acarol.* 5: 225-241.

Dicke, M. 1988. *Infochemicals in tritrophic interactions*, Ph.D. thesis. Agricultural University, Wageningen, Netherlands: 235.

Dimock, R.V. 1988. Photobiology and behaviour of the water mite genus *Unionicola*: a mini review. *Camp. Biochem. Physiol.* 91C: 193-197.

Dong, Y.C., Ran, R.B. and Xiang, J.Y. 1996. Biology of *Allothrombium ovatum* (Acari: Trombidiidae) and its controlling effect on *Aphis gossypii* (Homoptera: Aphididae). *Syst. Appl. Acarol.* 1: 35-40.

Dosse, G. 1961. Uber dic Bedeutung der Pollennahrung fur *Typhlodromus pyri* Scheuten (=tillae ond.) (Acari: Phytoseiidae). *Entomol. Exp. Appl.* 4: 191-195.

Dubinin, V.B. 1951. Feather mites (Ana1gesoidea). Part-I. *Introduction to the study. Fauna USSR Arachnida* 6 (5): 1–363.

Duso, C. and Camporese. 1991. Developmental times and oviposition rates of predatory mites *Typhloromus pyri* and *Amblyseius anderosoni* (Acari: Phytosciidae) reared on different foods. *Expt. Appl. Acarol.* 13: 117-128.

Edge, V.E. and James, D.G. 1982. Detection of cyhexatin resistance in two spotted mite, *Tetranychus urticae* koch. (Acarina: Tetranychidae) in Australia. *J. Austr. Entomol. Soc.* 21: 198.

Edwards, C. A. T. 1955. In: *Soil Zoology* (ed. Kevan, D. K.), McE. London: 412-416.

Elkins, N. Z. and Whitford, W. G. 1982. The role of micro arthropods and nematodes in decomposition in a semi arid ecosystem. *Oecologia,* 55: 303-310.

Ellingsen, I.J. 1974. *Comparison of active and quiescent protonymphs of the American house dust mites.* Ph.D. Dissertation, Ohio State University, Columbus: 82pp.

Ellingsen, I.J. 1978. Oxygen consumption in active and quiescent protonymphs of American house dust mites..*J. Insect. Physiol.* 24: 13-16.

Epsky, N.D., Walter, D.E. and Capinera, J.L. 1988. Potential role of nematophagous microorthropods as biotic mortality factors of entomogenous nematodes (Rhabditida: Skinernematidae, Heterochabditidae). *J. Econ. Entomol.* 81: 821-825.

Evans G.O. 1969. *Proc. 2nd Intl. Cong. Acarology,* Akademiai Kaido, Budapest, Hungary. 652pp.

Evans G.O. 1992. *Principles of Acarology*. CAB International. 563 pp.

Evans, G.O., Sheals, J.G. and MacFarlance, D. 1961. *Terrestrial Acari of the British Isles,* Adland and Sons, Bartholomew Press, Durking, UK.

Faccini, J.L.H. and Costa, A.L. 1992. Subclinical psoroptic otocariosis in Brazilian sheep with comments on a technique for mite collection. *Exp. Appl. Acarol.* 13: 227-229.

Faccini, J.L.H., Lignon, G.B. and Leite, R.C. 1987. Evaluation of an ear flushing technique as a post-mortem measure of infestation of *Raillietia allris* (Leidy) (Acari) in Cattle. *Exp. Appl. Acarol.* 3: 175-178.

Fain, A., Guerin, B. and Hart, B.J. 1988. *Acariens et Allergies. Allerbio, Varennes-en Argonne*, 179pp.

Fain, A. 1965. A review of the family Epidermoptidae Trouessart parasitic on the skin of birds (Acarina: Sarcoptiformes), Paleis Der Academien-Hertogsst raat; Brussel. Part-I: 1-176, Part-II: 5-144.

Fain, A. 1967. Le genre *Dermatophagoides* Bogdanov, 1864; son importance dansles allergies respiratiores et cutanees chezl[1] homme. *Acarologia.* 9: 179-225.

Faroni, L.R.D.A., Guedes, R.N.C. and Matioli, A.L. 2001. Effect of temperature on the development and population growth of *Acarophenax lacunatus* (Cross and Krantz) (Prostigmata: Acarophenaeidae) on *Rhyzopertha dominica* (F.) (Coleoptera: Bostrichidae). *Biocontrol Sci. Tech.* 11 (1-6): 5-12.

Fenley, W.R. 1966. Field observations of *Aedes ventrovittis* Dyar (Diptera: Culicidae) parasitized by *Panisopsis* sp. (Acarina: Thyasidae). *Mosq. News*, 26: 583-584.

Field, R.P. 1976. Integrated pest control in Victorian peach orchards: The role of *Typhlodromus occidentalis* Nesbitt (Acarina: Phytoseiidae). *Aust. J. Zool.* 24: 565-572.

Finney, D.J. 1971. *Probit Analysis*. Cambridge University Press. 333 p.

Fitzgerald, J.D. and Solomon, M.G. 1992. Field selection of the predatory mite *Typhlodromus pyri* for resistance to pyrethroids. *Proc. 1992 Brit. Crop Protec. Conf.*: 1199-1204.

Flechtmann, C.H.W., Guerrero, B.J.M., Arroyave, J.A. and Constantino, C.L.L.M. 1990. A little known mode of dispersal of *Polyphagotersonemus latus* (Banks) *Int. J. Acarol.* 16: 181-182.

Fletcher, M.G. and Axtell, R.C. 1991. Susceptibility of northern fowl mite, *Ornithonyssus sylviarum* (Acarine: Macronyssidae), and chicken mite, *Dermanyssus gallinae* (Acarina: Dermanyssidae), to selected acaricides. *Exp. Appl. Acarol.* 13: 137-142.

Fotedar, R. 1981. Effect of host plant nitrogen content and age Dolichos lablab foliage on the fecundity of *Tetranychus cinnabarinus*. In: *Contributions to Acarology in India* (ed. Channabasavanna GP), Acarological Society of India, Bangalore: 58-61.

Foulk, J.D. and Mattysse, J.G. 1964. A new toxicological text method for hematophagous mites. *J. Econ. Entomol*. 57: 602-604.

Frazer, B.D. and Nelson, C. 1981. Note on the occurrence of predatory *Anystis* mites (Acari: Anystidae) in SW British Columbia. *J. Entomol. Soc. British Columbia* 78: 46.

Fries, I., Aarhus, A., Hansen, H. and Korpela, S. 1991. Development of early infestations by the mite *Varroa jacobsoni* in honeybee (*Apis mellifera*) colonies in cold climates. *Exp. App. Acarol*. 11: 205-214.

Fries, I., Camazine, S. and Sneyd, J. 1994. Population dynamics of *Varroa jacobsoni*: a model and a review. *Bee Wld*. 75 (1): 5-28.

Fuchs, S. 1986. Conditions when reproduction of *Varroa* in worker cells is advantageous. The relationship between mite reproduction, mite population size, number of worker cells and number of drone cells. *Apidologie*.17: 368-371.

Fuchs, S. 1990. Drone brood cell preference of *Varroa jacobsoni* Oud. in colonies of *Apis mellifera carnica*. *Apidologie*. 21: 193–199.

Furumizo, R.T.1975. Laboratory observations on the life history and biology of the American hose dust mite, *Dermatophagoides farinae* (Acarina: Pyroglyphidae). *Calif. Vector Views*. 22: 49-60.

Gabriel, M., Cunnington, A.M., Allan, W.G.L., Pickering, C.A.C. and Wraith, D.G. 1982. Mite allergy in Hong Kong. *Clin. Allergy* 12: 157-171.

Garg, R. and Kashyap, N.P. 1998. Mites and other enemies of honeybees in India. In: *Perspectives in Indian Apiculture* (ed. Mishra, R.C.): 264-303.

Garg, R. and Sharma, O.P. 1988. Efficacy of chlorobenzilate and menthol against *Tropilaelaps clareae* Delfinado and Baker (Laelaptinae:Acarina) *in Apis mellifera* L. *Indian Bee J*. 50 (3): 65-66.

Garg, R. and Sharma, O.P. 1994. Biology of *Tropilaelaps clareae* Delfinado and Baker (Laelaptinae:Acarina) an ectoparasite of *Apis mellifera* in tropical Asia. In: *Nat. Sem. Cum Workshop Honeybee Dis.Diagn. Manage.*, Bangalore:21.

Garg, R., Sharma, O.P. and Dogra, G.S. 1984. Formic acid: an effective acaricides against *Tropilaelaps clareae* Delfined and Baker (Laelapidae: Acarina) and its effect on brood and longevity of honeybees. *Am. Bee J.* 124: 736-38.

Gauchat, C.A. 1972. A note on *Podapolipoides grassi* Berlese (Acarina: Podapolipidae), a parasite of *Chortoicetes terminifera* Walker, the Australian plague locust. *J. Aust. Entomol. Soc.* 11: 259.

Gencsoylu, I., Liu, W., Usmani, K.A. and Knowles, C.O. 1998. Toxicity of acaricides to the bulb mite, *Rhizoglyphus echiropus* Acari: Acaridae). *Exp. App. Acarol.* 22: 343-351.

Gentile, A.G., Wobb, R.E. and Stoner, A.K. 1969. *Lycopersicon* and *Solanum* spp. resistant to the canine and the two spotted spider mite. *J. Econ. Entomol.*62: 834-836.

Gerson, U. and Cohen, E. 1989. Resurgences of spides mites (Acari: Tetranychidae) induced by synthetic pyrethoids. *Exp. App. Acarol.* 6: 29-46.

Gerson, U. and Blumberg, D. 1969. Biological notes on the mite, *Saniosulus nudus*. *J. Econ. Entomol.* 62: 729-730.

Gerson, U. and Schneider, R. 1981. Laboratory and field studies on the mite *Hemisarcoptes coccophagus* Meyer (Astigmata: Hemisarcoptidae), a natural enemy of armoured scale insects *Acarologia* 22: 199-208.

Gerson, U. and Smiley, R.L. 1990. *Acarine biocontrol agents*: An illustrated key and Manual. Chapmaa and Hall, London: 174pp.

Ghai, S. 1964. Mites. In: *Entomlogy in India. Ent. Soc. India*: 384-496.

Ghai, S. 1976. Mites associated with stored food in India. *Bull. Grain Tech.* 14(2): 134-149.

Ghosh, S.K. 2001. Prospects of employing predatory mites for the management of phytophagous mites. In: *Biodiversity and Management of Phytophagous Mites* (eds. Rabindra, R.J., Karuppuchamy, P., Umapathy, G., Chinniah, C. and Balasubramani, V.): 181–188.

Glass, E.V., Yoder. J.A. and Needhan, G.R. 1998. Clustering reduces water loss by adult American house dust mites, *Dermatophagoides farinae* (Acari: Pyroglyphidae). *Exp. Appl. Acarol.* 22(1): 31-38.

Glinski, Z. and Jarosz, J. 1992. *Varroa jacobsoni* as a carrier of bacterial infections to a recipient bee host. *Apidologie* 23: 25-31.

Goedon RD. 1983. Critique and revision of Harris' scoring system for selection of insect agents in biological control of weeds. *Prot Ecol.* 5: 287-301.

Goldschmidt, T. 2001. *Die Wassermilbenfauna Costa Ricas (Hydrachnidia, Acari) – Systematik, Ökologie, Diversität und Zoogeographie.* Dissertation, Westfälische Wilhelms-Universität Münster: 425 pp.

Goldschmidt, T. 2002. The Biodiversity of Neotropical Water Mites. In: *Acarid Phylogeny and Evolution: Adaptation in Mites and Ticks.* (eds. Bernini, F., Nannelli, R., Nuzzacis, G. and Delillo, E.). In: *Proc. 4th Sym. EURAAC* 24-29 July 2000, Siena, Italy. Kluwer Academic Publishers. Dortrecht, Boston, London. 91-99 pp.

Goldschmidt, T., Albert, G. and Meyer, E.I. 1999. Presence of acetabula-like structures on the coxae of the neotropical water mite genus *Neotyrrellia* (Tyrrellinae, Limnesiidae, Prostigmata). In: *Ecology and Evolution of the Acari* (eds. Bruin, J., Vander Geest, L.P.S. and Sabelis, M.W.) Kluwer Academic Publishers, Dordrecht, Boston, London: 491-497.

Gope, D. and Das, S.C. 1992. Biology and effect of pruning and skiffing on the distribution of the scarlet mite, *Brevipalpus phoenicis* (Geijskes) on tea in North East India. *Two and a Bud* 39: 23.

Goyal, N.P. and Kandoria, J.L. 1988. Acarine disease of honey bees and its control. *Indian Bee J.* 50:85-86.

Grafton, C.E., Yuling, Q., and Quagang, Y.L. 1995. Manipulation of the predaceous mite, *Euseius tularensis* (Acari: Phytoseiidae) with pruning for citrus thrips control. In: *International Conf. Thysanoptera* (eds. Parker, B.L. *et al.*). Thrips biology and management: 251-254.

Graham, W.M. 1970. Warehouse ecology studies of bagged maize in Kenya – II. Ecological observations of an infestation by *Ephestia*

(*Cendra*) *cautella* (Walker) (Lepidoptera, phyeitidae) *J. Stored Prod. Res.* 6: 157-167.

Grandjean, F. 1935. Observations sur les Acariens. *Bull. Mus. Nat.Hist. Natur. Paris Ser.* 2 7: 201-208.

Grandjean, F. 1949. Obsrvation et conservation des tres petito Arthropodes. *Bulletin du Museum d'Histoire Naturella*, Paris, 21: 363-370.

Grandjean, F. 1972-76. *Oeuvres Acarologiques Completes Series* (Vol 1-7), Lochem, The Hague.

Green, W.F. and Woodcock, A.J. 1978. *Tyrophagus putrescentiae* an allergenically important mite. *Clin Allergy*. 8: 135-144.

Griffiths, D.A., Hodson, A.C. and Christensen, C.M. 1959. Grain storage fungi associated with mites. *J. Econ. Entomol.* 52: 514-518.

Griffiths, D.A. 1966. Nutrition as a factor influencing hypopus formation in the *Acarus siro* species complex (Acarina, Acaridae). *J. Stored Prod. Res.* 1: 325-340.

Grobov, O.F. 1976. Varroasis in Bees. *Symposium Sofia*: 46-90.

Gulati, R. 1998. Inhibitory action of neem products on *Tyrophagus putrescentiae* Schrank (Acarina: Acaridae) in wheat during storage. *Ann. Agri–Bio Res.*: 227-230.

Gulati, R. 2002. Bioefficacy of *Curcuma longa* L. oil against *Tyrophagus putrescentiae* Schrank and *Suidasia nesbitti* Hughes in wheat. *Pest Man. Econ. Zool.* 10(2): 115-119.

Gulati, R. 2003. Acarines as bio control agents of insect pests. In: *Emerging Trends in Biological Control of Insect Pests and Weeds* (eds. Chhillar BS, Kaushik HD, Malik VK, Ombir) CCS HAU, Hisar: 146-153.

Gulati, R. 2004. Incidence of *Tetranychus cinnabarinus* (Boisd.) infestation in different varieties of *Abelmoschus esculentus* L. *Ann.Pl.Protec.Sci.* 12 (1): 45-47.

Gulati, R. and Mathur, S. 1995. Effect of *Eucalyptus, Mentha* leaves and *Curcuma* rhizomes on *Tyrophagus putrescentiae* (Schrank) in wheat. *Exp.Appl. Acarol.* 19: 511-519.

Gulati, R and Mathur, S. 1998. Olfactory responses of *Tyrophagus putrescentiae* (Schrank) towards *Cajanus cajani*. *Bull. Life Sci.* 8: 53-61.

Gulati, R., Singh, M., Punia, A. and Sareen, B. 1999. Estimation of germination losses in four pulses from infestation by *Suidasia nesbitti. J. Acarol.* 14: 112-115.

Gulati, R., Walia, K.K. and Rani, R. 2005. Acarines as bio control agents of plant parasitic nematodes. In: *Crop Protection: Management Stratergies* (ed. Prasad, V): 526-534.

Gupta, S.K.1985. *Handbook–Plant mites of India*. ZSI, Calcutta. 520pp.

Gupta, S.K. 1986. *Fauna of India. Acari: Mesostigmata. Family Phytoseiidae*. ZSI, Calcutta. 350 pp.

Gupta, S.K. 2003. Taxonomy and bioecology of predatory mites of India with special emphasis on phytoseiidae. In: *Mites, their identification and management* (ed. Yadav, P.R., Chauhan, R., Putatunda, B.N. and Chhillar, B.S.) Centre of Advanced Studies, Department of Entomology, CCSHAU, Hisar (India): 168-181.

Gupta, S.K. and Gupta, A. 1992. In: *Man, mites and environment*: 146-154.

Gupta, S.K. and Gupta, A. 1999. Progress of taxonomic research on Indian mites upto the end of twentieth century and prospects of research in the next millennium. *J. Acarol.* 15(1&2): 80-83.

Gupta, S.K., Sidhu, A.S. and Singh, G. 1971. Occurrence of a tenuipalpid mite on citrus in the Punjab and its control. *Indian J. Ent.* 33(1): 30-33.

Gupta, S.K.and Chatterjee, K. 1997. In: *State fauna Scr. 6, Fauna of Delhi*, 485-522.

Guthrie, W.D., Russel, W.H., Bing, J.W. and Dicker, F.F. 1991. Registration of B-96 germplasm line of maize. *Crop Sci.* 31: 231-240.

Hadam, D., Ali Niazee, M.T. and Croft, B.A. 1986. Phytoseiid mites (Parasiteformes: Phytoseiidae) of major crops in Willamette Valley, Oregon, and pesticide resistance in *Typhlodromus pyri* Schenten. *Environ. Entomol.* 15: 1255-1263.

Hage-Hamsten, M.V. and Johansson, S.G.O. 1992. Storage mites. *Exp. Appl. Acarol.* 16:117-128.

Hagley, E.A.C and Simpson, C.M. 1983. Effect of insecticides on predators of the pear psylla, *Psylla pyricola* (Hemiptera: Psyllidae) in Ontario. *Can Entomol.* 115: 1409-14.

Haines, C.P. 1981. Laboratory studies on the role of an egg predator, *Blattisocius tarsalis* (Berlese) (Acari: Ascidae), in relation to the natural control of *Ephestion cantella* (walker) (Lepidoptera: Pyralidae) in warehouses, *Bull. Entomol. Res.* 71: 555-574.

Hall, R.D., Townsend, L.H.Jr and Turner, E.C.Jr. 1978. Laboratory and field tests to compare the effectiveness of organophosphorus, carbomates and synthetic pyrethroid acaricides against northern fowl mites. *J. Econ. Entomol.* 71: 316-318.

Hamamura, T., Shinkaji, N. and Aghihara, W. 1978. Studies on the low temperature storage of *Phytoseiulus persimilis* Athias-Henriot (Acarina. Phytoseiidae). *Bull. Fruit Tree Res. Stu.* Series (Akitsu) 2: 83-90.

Hansen, L.S. 1988. Control of *Thrips tabaci* (Thysanoptera: Thripidae) on glass house cucumber using large introductions of predatory mites *Amblyseius barkeri* (Acarina: Phytoseiidae). *Entomophaga* 33: 33-42.

Haq, M.A. and Ramni, N. 2003. Sampling, extraction and identification of oribatid mites. In: *Mites, their identification and management* (eds. Yadav, P.R., Chauhan, R., Putatunda, B.N. and Chhillar, B.S.), Dept. of Entomology, CCSHAU, Hisar: 195-212.

Haq, M.A., Ramni, N. and Prakashan, K. 1999. Oribated mites as transmitting agents of cestodes in Kerala. *J. Acarol.* 15(1&2): 68-72.

Harris, P. 1973. The selection of effective agents for the biological control of weeds. *Can. Entomol.* 105: 1495-1503.

Hart, B.J. and Fain, A. 1987. A new technique for isolation of mites exploiting the difference between ethnol and saturated NaCl: qualitative and quantitative studies. *Acarologia.* 28: 251-254.

Hart, B.J. and Fain, A. 1988. Morphological and biological studies of medically important house dust mites. *Acarologia.* 19: 285-295.

Hart, B.J. and LeMerdy, L. 1988. Human dander-free house dust mite extracts. In: *Mite Allergy-A worldwide problem* (eds De Weck, A.L. and Todt, A.) UCB Institute of Allergy, Brussels: 47-50.

Hart, B.J. and Whitehead, L. 1990. Ecology of house dust mites in Oxfordshire. *Clin. Exp. Allergy.* 20: 203-209.

Harvey, M.S. 1990. Pezidae, a new freshwater mite family from Australia (Acarina: Halacaroidea). *Invertebr. Taxon.* 3: 771-781.

Harvey, M.S. 1998. *The Australian Water Mites: a Guide to Families and Genera*. CSIRO Publishing, Collingwood, Victoria.

Harvey, T.L., Martin, T.J. and Seifers, D.L. 1995. Survival of five wheat curl mite, *Aceria tosichilla* Keifer (Acari: Eriophyidae), strains on mite resistant wheat. *Exp. App. Acarol.* 19: 459-463.

Hariyappa, A.S.and Kulkarni, K.A. 1989. *J. Biol. Contr.* 3(1):31-32.

Hassan, S.A., Bigler, F., Bogenschutz, H., Bolher, E. and Brun, J. 1988. Results of the fourth joint pesticide testing programme carried out by the IOBC/WPRS–Working Group "Pesticides and Beneficial Arthropods". *J. Appl. Entomol.* 105: 321-329.

Hay, D.B., Hart, B.J., Pearce, R.B., Kozakiewicz, Z. and Douglas, A.E. 1992. How relevant are house dust mite-fungal interactions in laboratory culture to the natural dust system? *Exp. Appl. Acarol.* 16: 37-47.

Haydak, M.H. 1936. A food for rearing laboratory insect. *J. Econ. Ent.* 29: 1026.

Helle, W. and Sabelis, M.W. 1985. *Spider mites, their Biology, Natural Enemies and Control. World Crop pests.* Vol. IB Elsevier, Amsterdam: 458pp.

Henderson, C.F. and McBurnie, H.V. 1943. Sampling techniques for determining populations of the red citrus mite and its predators. *US Deptt. Agric. Circular* 671: 1-11.

Henderson, C.F. and McBurnie, H.V. 1954. Sampling Techniques for Determining Populations of Citrus Red Mite and its Predators. US Department of Agriculture.

Henderson, C.P. 1960. A sampling technique for estimating populations of small arthropods in soil and vegetation. *J. Econ. Entomol.* 53: 115-121.

Herbert, H. J. 1981. Biology, life tables, and innate capacity for increase of the two-spotted spider mite, *Tetranychus urticae* (Acarina: Tetranychidae). *Can Entomol.* 113: 371-78.

Herren, H.R., Gilstrap, F.E., Morrison, R.K. and Bouse, L.F. 1987. Release of predatory mites (Acari: Phytoseiidae) by aircraft for the biological control of spider mites (Acari: Tetranychidae) infesting corn. *J. Econ. Entomol.* 40: 906-910.

Herron, G.A., Beattie, G.A.C., Kallianpur, A. and Barchia, I. 1998. A potter spray tower bioassay of two petroleum spray oils against adult female *Panonychus ulmi* (Koch) and *Tetranychus urticae* Koch. (Acari: Tetranychidae). *Exp. Appl. Acarol.* 22: 553-558.

Heymann, P.W., Chapman, M.D., Aalberse, R.C., Fox, J.W. and Platts-Mills, T.A.E. 1989. Antigenic and structural analysis of group II allergen (*Der f* II and *Der p* II) from house dust mites (*Dermatophagoides* spp.). *J. Allergy Clin. Immunol.* 83: 1055-1067.

Hill, R.L. and Stone, C. 1985. Spider mites as control agents for weeds. In: *Spider mites, their biology, natural enemies and control* (eds. Helle, W. and Sabelis, M.W): 443-448.

Ho, C. C. and Chen, W. H. 1992. Control of phytoseiids in a spider mite mass-rearing system (Acari: Phytoseiidae, Tetranychidae). *Exp. Appl. Acarol.* 13: 287-293.

Hoppe, H., Ritter, W. and Stephen, E.W.C. 1989. The control of parasitic bee mites: *Varroa jacobsoni, Acaropis woodi* and *Tropilaelaps clareae* with formic acid. *Am. Bee J.* 129: 739-42.

Hosamani, R.K., Sharma, S.K. and Gulati, R. 2005. Effect of *Tropilaelaps clareae* Delfinado and Baker (Mesostigmata: Laelapidae) Infestation on Colony Buildup and Stores in *Apis mellifera* L. Colonies. *Korean J. Apicult.* 20(1): 35-38.

Hosamani, R.K., Sharma, S.K. and Gulati, R. 2006. Pest potential of *Tropilaelaps clareae* Delfinado and Baker (Mesostigmata: Laelapidae) on *Apis mellifera* L. colonies in Hisar, India. *Honeybee Sci*: in press.

House, H.L. 1980. Artificial diets for the adult parasite *Itoplectis conquisitor* (Hymenoptera: lchneumonidae). *Can. Entomol.* 112: 315-320.

Howard, L.O. 1930. *A History of Applied Entomology (Somewhat Ancedotal)*. Smithsonian Misc. coll. Publ. No. 3065:564pp.

Hoy, M.A. 1976. Genetic improvement of insects: Fact or fantasy. *Environ. Entomol.* 5: 833-839.

Hoy, M.A. 1983. Opportunities for genetic improvement of mites as biological control agents. In: *Biological Control of Pests by Mites* (eds. Cunningham, G.L., Hoy, M.A. and Knudson, L.), Division of Agri. and Natural Resources Univ. California, USA: 111-114.

Hoy, M.A. 1985. Recent advances in genetics and genetic improvement of the phytoseiidae. *Ann. Rev. Entomol.* 3: 345-370.

Hoy, M.A., Flaherty, O., Peacock, W. and Culver, O. 1979. Vineyard and laboratory evaluations of methomyl, dimethoate and permethrin for a grape pest management program in the San Joaquin Valley of California. *J. Econ. Entomol.* 72: 250-255.

Hoy, M.A. and Knop, N.E. 1981. Selection for and genetic analysis of permethrin resistance in *Metaseiulus occidentalis*: genetic improvement of a biological control agent. *Entomol. Exp. App.* 30: 10-18.

Hoy, M.A. and Smilanick, J.M. 1981. Non-random prey location by the phytoseiid predator, *Metaseiulus occidentalis* differential response to several spider mite species. *Entomol. Exp. Appl.* 29: 241-253.

Hoy, M.A., Castro, D. and Cahul, D. 1982. Two methods for large scale production of pesticide resistant strains of the spider mite predator, *Metaseiulus occidentalis* (Nesbitt) (Acarina: Phytoseiidae*). Z. Angew. Entomol.* 94: 1-9.

Hoyt, S.C., Westigard, P.H. and Burts, E.C. 1978. Effects of two synthetic pyrethroid insecticides on the codling moth, pear psylla and various mite species in North West apple and pear orchards. *J. Econ. Entomol.* 71: 431-434.

Huffaker, C.B., Simmonds, F.J. and Laing, J.E. 1976. The theoretical and empirical basis of biological control. *In: The Theory and Practice of Biological Control* (eds. Huffaker, C.B. and Messenger, P.S.) Academic Press, New York: 41-78.

Hughes, A.M. 1976. *Mites of stored food and houses*. Publ. By Her Majesty's Stationary Office, London: 400pp.

Hull, L.A. and Beer, E.H. 1989. Validation of injury thresholds for European red mite (Acari: Tetranychidae) on 'Yorking' and 'Delicious' apple. *J. Econ. Entomol.* 83: 2026-2031.

Hunter, WO and Hinds, W.E. 1904. The Mexican cotton boll weevil. *USDA Div. Entomol.*. Bull No. 45: 116pp.

Hurtado, I. and Parini, M. 1987. House dust mites in carcas, Venezuela. *Ann. Allergy* 59:128-130.

Iation, G., Cook, C.M., Stamon, G. and Lanaras, T. 1995. Chlorophyll fluorescence and leaf chlorophyll content of bean leaves injured by spider mites (Acari: Tetranychidae). *Exp. App. Acarol.* 19: 581-591.

Ifantidis, M.D. 1988. Some aspects of the process of *Varroa jacobsoni* mite entrance into honeybee (*Apis mellifera*) brood cells. *Apidologie*. 19: 387-396.

Ifantidis, M.D. 1990. Reexamination of reproduction parameters of the mite *Varroa jacobsoni* Oudemans. In: *Recent Research on Bee Pathology* (ed. Ritter W.): 20-26.

Ifantidis, M..D. and Rosenkranz, P. 1988. Reproduktion der Bienenmilbe *Varroa jacobsoni* (Acari: Varroidae). *Entomol. J.* 14: 111-122.

Ignatowicz, S. 1980. Effect of inorganic salts on biology and development of acarid mites. 4. Effect of calcium phosphate surplus in food on fecundity, life span and development of copra mite, *Tyrophagus putrescentiae* (Acarina: Acaridae). *Pot. Pismo. Entomol.* 50(2): 289-298.

Ignatowicz, S. and Pankiewicz-Nowicka, D. 1980. Effect of inorganic salts on biology and development of acarid mites. 6. Effect of tricalcium phosphate and egg viability of flour mite, *Acarus siro* (Acarina: Acaridae). *Pol. Pismo Entomol.* 50(4): 110-113.

Imbriani, L. and Mankau, R. 1983. Studies on *Lasioseius scapulatus*, a mesostigmatid mite predaceous on nematodes. *J. Nematol.* 15: 523-528.

Inserra, R.N. and Davis, D.W. 1983. *Hypospis* nr. *aculenfer:* a mite predaceous on root knot and cyst nematodes. *J. Nematol.* 15: 324-325.

Ishibashi, N., Young, F., Nakashima, M., Abiru, C. and Horaguchi, N. 1987. Effect of application of DD-136 on silkworm, *Bombyx mori*, predatory insect, *Agriosphodrus dohrni*, parasitoid, *Trichomalus apanteloctenus*, soil mites and other non-target soil arthropods, with brief notes on feeding behavior and predatory pressure of soil mites, tardigrades and predatory nematodes on DD-136 nematodes. In: *Recent advances in biological control of insect pests by entomogenous nematodes in Japan*, Ministry of Education, culture and Science, Sago, Japan: 158-164.

Ito, Y. 1971. Predation by manure inhabiting mesostigmatids (Acarina: Mesostigmata) in some free living nematodes. *Appl. Ent. Zool.*, 6: 51-56.

Iversen, M., Korsgaard, J., Hallas, T. and Dahl, R. 1990. Mite allergy and exposure to storage mites and house dust mites in farmers. *Clin. Exp. Allergy* 20: 211-219.

Jacobson, R.J., Croft, P. and Fenlon, J. 2001. Suppressing establishment of *Frankliniella occidentalis* Pergande (Thysanoptera: Thripidae) in cucumber crops by prophylactic release of *Amblyseius cucumeris* Oudemans (Acarina: Phytoseiidae) 11 (1-6): 27-34.

Jacot, A.P. 1932. A knock-down "Berlese Funnel". *Science* 76: 299-300.

Jagadish, P.S. and Nageshchandra, B.K. 1979. *Abst. I. All India Symp. Acarology*: 49.

Jagadish P. S. and Nageshchandra, B.K. 1982. *Acar. Newsl.* 11: 9-10.

Jalil, M. and Rodriguez, J.G. 1970a. Biology of and odour perception by *Fuscuropoda vegetans* (Acarina: Uropodidae), a predator of the housefly. *Ann. Entomol Soc. Am.* 63: 935-938.

Jalil, M. and Rodriquez, J.G. 1970b. Studies of behaviour of *Macrocheles muscadomesticae* (Acarina: Macrochelidae). *Ann. Entomol. Soc. Am.* 63: 733-738.

James, D. G. 1989. Influence of diet on development, survival and oviposition in an Australian phytoseiid, *Amblyseius victoriensis* (Acari: Phytoseiidae). *Exp. Appl. Acarol.* 6: 1-10.

Janarthan, R., Navaneethan, G., Subramanian, K.S. and Samuel, G.S. 1972. A method for assessment of eriophyd mites on pigeonpea leaves. *Madras Agric. J.* 59: 437.

Jedlickova, J. 1997. Modification of a leaf washing apparatus for the recovery of mites. *Exp. Appl. Acarol.* 21: 273-277.

Jeffrey, I.G. 1976. A survey of the mite fauna of Scottish farms. *J. Stored Prod. Res.* 12: 149-156.

Jensen, W.A. 1962. *Botanical histochemistry*, W.H. Freeman and Co., San Fransisco: 408 pp.

Jeppson, L.R., Keifer, H.H. and Baker, E.W. 1975. *Mites Injurious to Plants*. Univ. of Calif. Press, Berkeley: 614 pp.

Johansen, D.A. 1940. *Plant Micro Technique.* McGraw-Hill Publications, New York: 523 pp.

Johansson, S.G.O. and Van Hage-Hamsten, M. 1990. Berufrbedingle Atemwegs Allergie bei Landwirten insberondere-Allergien, die durch Yorrat smilben verursacht werden. *Allergologie* 13: 344-348.

Johnson, D.T. and Croft, B.A. 1976. Laboratory study of dispersal behaviour of *Amblyseius fallacis* (Acarina: Phytoseiidae). *Ann. Ent. Soc. Am.* 69:1019-1023.

Johnson, D.T. and Croft, B.A. 1979. Factors affecting the dispersal of *Amblyseius fallacis* in an apple tree ecosystem. *Rec. Adv. Acarol.* 1: 477-483.

Jose, V.T., Shah, A. H. and Patel, C.B. 1989. In: *Progress in Acarology.* 2: 357-360.

Kanta, V. 1971. *Biological Studies on the mite, Eutetranychus orientalis (Klein), a pest of citrus trees in Punjab.* M.Sc. Thesis, PAU, Ludhiana: 97pp.

Kantaratanakul, S. and Rodriguez, J.G. 1979. Nutritional studies in *Tetranychus urticae* Koch (Acarina Tetranychidae). 1. Development of a chemically defined diet. *Int. J. Acarol.* 5 (2): 83-92.

Kapil, R.P. and Aggarwal, K. 1987. Concurrent parasitization of *Apis florea* by *T. clareae* and *Euvarroa sinhai Exptl. Appl. Acarol.* 3 (3):267-269.

Kapil, R.P. and Aggarwal, K. 1989. Observations on reproduction and seasonal population trends of *Euvarroa sinhai* (Mesostigrnata:Varroidae) in India. In: *Progress in Acarology,* Oxford and IBH Publ. Co. Pvt. Ltd, New Delhi 2: 277-281.

Kapil, R.P. and Bhanot, J.P. 1971. *Acaropsis docta* (Berlese): a predaceous mite of *Trogoderma granarium* (Everts.). *Indian J. Ent.* 33 (4): 457.

Kapil, R.P and Bhanot, J.P. 1973. Feeding behaviour of the predatory mite, *Acaropsis docta* (Berlese). *J. Stored Prod. Res.* 9: 1-6.

Karban, R., English-Locb, G., Walker, M.A. and Thaler, J. 1995. Abundance of phytoseiid mites on *Vitis* species: effects of leaf hairs, domatia, prey abundance and plant phylogeny. *Exp. App. Acarol.* 19: 189-197.

Karg, W. 1961. Okologische Untasuchungen von edaphischen gamasiden (Acari: Parasitiformes). *Pedobiologia* 1: 58-98.

Karg, W. 1971. Acari (Acarina), Milben, Unterordnung Anactinochaeta (Parasitiformes). Die freilebenden Gamasina (Gamasides), Raubmilben. In: *Die Tierwelt Deutschlands und der angrenzenden Meeresteile,* 59. *Teil Gustav Fischer Verlag,jena,* 475.

Karg, W. 1983. Verbreitung and Bedeutung Von Vau-bonilben der Cohors Gamasina als Antagonisten Von. Nematoden. *Pedobiologia* 25: 419-432.

Karuppuchamy, P. and Mohanasundaram, M. 1987. Bioecology and control of chilli muranai mite, *Polyphagotarsonemus latus* (Banks) (Tarsonemidae: Acari). *Indian J. Pl. Prot.* 15: 1-4.

Keil, H. 1983. Okofaunistosche Untersuchungen der Hausstaubmilben in Hambung. Entomol. Mit. *Zool. Mus. Hambung*. 7: 343-386.

Kempson, D., Lloyd, M. and Ghelardi, R. 1963. A new extractor for woodland litter. *Pedobiologia*. 3: 1-21.

Kerfoot, W.C. 1982. A question of taste: crypsis and warning coloration in freshwater zooplankton communities. *Ecology* 63: 538-554.

Khajuria, D.R. 2000. Efficacy of various pesticides against European red mite, *Panonychus ulmi* (Koch) on apple. *Indian J. Pl. Prot.* 28(1): 97-99.

Khanna, A. 1991. *Bionomics of some important mite pests of vegetable crops at Hisar.* M.Sc. Theses, CCSHAU, Hisar.

Khanna, K.L. and Ramanathan, K.R. 1947. Studies on the association of plant characters and pest incidence. I. Structure of leaf surface and mite attack in sugarcane. *Proc. Nat. Inst. Sci. India.* 13: 327-329.

King, E.J. and Morrison, R.K. 1984. Some systems for production of eight entomophagous arthropods. In: *Advances and challenges in Insect Rearing* (eds. King, E.G. and Leppla, N.C.). Agric. Res. Service (Southern Region). US Dept. Agric. New Orleans: 206-222.

Kinn, D. N. 1983. The life cycle of *Proctolaelaps dendroctoniLindquist* and Hunter (Acari: Ascidae); a mite associated with pine bark beetles.lnternat. *J. Acaro1.* 9: 205.

Kinn, D.N. 1984. Life cycle of *Dendrolaelaps neodisetus* (Mesostigmata: Digamasellidae), a nematophagous mite associated with pine bark beetles (Coleoptera: Scolytidae). *Environ. Entomo1.* 13: 1141-1144.

Knop, N.F. and Hoy, M.A. 1983. Tydeid mites in vineyards. *Calif. Agic.* 37: 16-18.

Knowles, C.O., Errampalli, D.D. and El-Sayed, G.N. 1988. Comparative toxicities of selected pesticides to bulb mite (Acari: Acaridae) and six spotted spider mite (Acari: Tetranychidae). *J. Econ. Entomol.* 81: 1586-1591.

Knutte, W. 1987. Genetic variability and ecological adaptability of hypopus formation in a stored product mite. *Exp. App. Acarol.* 3: 21-32.

Kohli, R. 1993. *Studies on feeding and reproductive behavior of an acarid mite, Tyrophagus putrescentiae (Schrank) (Acarina: Acaridae)*. Ph.D. Thesis, CCSHAU, Hisar.

Kohli, R. and Mathur, S. 1993. Life processes of *Tyrophagus putrescentiae* Schrank (Acarina: Acaridae) as influenced by photoperiod. *Crop Res.* 6: 311-316.

Kohli, R. and Mathur, S. 1994. Feeding behavior of an acarid mite, *Tyrophagus putrescentiae* (Schrank). *Crop Res.* 7(3): 466-472.

Konishi, E. and Uehara, K. 1990. Enzyme linked immunosorbent assay for quantifying antigens of *Dermaophagoides farinae* and *D. Pteromyssinus* (Acari: Pyroglyphidae) in house dust samples. *J. Med. Entomol.* 27: 993-998.

Konishi, E. and Uehara, K. 1994. Antigen levels of *Dermatophagoides* mites (Acari: Pyroglyphidae) in dust samples collected in homes of allergic patients. *J. Med. Entomol.* 31: 394-399.

Konishi, E. and Uehara, K. 1995. Distribution of *Dermotophagoides* mite (Acari: Pyroglyphidae) antigens in homes of allergic patients in Japan. *Exp. App.Acarol.* 19: 275-286.

Korsgaard, J. 1983. House dust mites and absolute indoor humidity. *Allergy.* 38: 85-92.

Koschier, E. 1997. Influence of different pruning measures on the population dynamics of European red mite (*Panonychus ulmi* Koch) and apple rust mite, *Aculus schlechtendali* (Nalepa) on apple trees. Mitteilungen Klosternenburg-Rebe-Und-Wein Obstban-Und-Fruchteverwertugig 47: 44-45.

Kovalev, O.V. 1973. Modern outlooks of biological control of weed plants in the USSR and the international phytophagous exchange. In: *Proc. 2nd Int. Symp. Biol. Contr. Weeds* (ed. Dunn, P.H.), Commonwealth Agricultural Burreaux: 166-172.

Krantz, G.W. 1978. *A Manual of Acarology*. 2nd edition. Oregon State University Book Stores, Corvallis, Oregon.

Krebs, J.R. 1978. Optimal foraging: decision rules for predators. In: *Behavioural Ecology: An evolutionary approach* (eds. Krebs JR, Davies NB), Blackwell Science Publications, Oxford: 23-63.

Krezeczkowski, K. 1961. Researches on the occurrence of the flour mite (*Tyrophagus noxins* Zachiw, Tyroglyphidae, Acarina) and its food preferences. *Prace namk. Inst. Ochr. Rost.* 3: 101-127.

Krishnamoorthy, A. 1983. *Entomon.* 8(3): 229-234.

Krishnamoorthy, A. 1988. *J. Biol Contr.* 2: 53-55.

Krishnamoorthy, A. 1990. *Acar Newsl.* 17: 18-28.

Kshirsagar, K.K., Mittal, M.C., Mahindre, D.S., Salvi, S.R. and Chauhan, R.M. 1980. Seasonal fluctuations in the incidence of acarine disease in bees and colony losses in India. *Proc. 2nd Intern. Conf. Apic. Trop. climates,* New Oelhi: 442-453.

Kumari, M. and Sadana, G.L 1995a. *J. Acarol.* 13: 55-62.

Kumari, M. and Sadana; G.L.1995b. *J. Acarol.* 13: 69-74.

Kumud, 1987. *Analysis of the interactions between some stored product mites and storage fungi in a micro-cosm*. Ph.D. Thesis, CCSHAU, Hisar: 52 pp.

Kumud, Mathur, R.B. and Mathur, S. 1988. A preliminary report on the occurrence of house dust mites from Hisar, the subtropical region of Haryana, India. *Progress in Acarology,* 1: 157-162.

Kuwahara, M. 1986. Resistance of the bulb mite, *Rhizoglyphus robini* Claparede (Acarina: Acaridae) to insecticides. I Resistance patterns to organophosphorus insecticides. *Jpn. J. App. Entomol. Zool.* 30: 290-295.

Kuwahara, M., Takai, M. and Fujimoto, K. 1985. Simple rearing of the bulb mite, *Rhizoglyphus robini* and test methods of acaricides *Syokubutsu-bocki Pl. Protec.* 39: 68-70.

Laasko, J. and Setala, H. 1999. Population and ecosystem level effects of predation on microbial feeding nematodes. *Oecologia* 120(2): 279-286.

Ladell, W.R.S. 1936. A new apparatus for separating insects and other arthropods from the soil. *Ann. Appl. Biol.* 23: 862-879.

Laigo, F.M. and Morse, R.A. 1968. The mite, *Tropilaelaps clareae* in *Apis dorsata* colonies in the Philippines. *Bee World.* 49: 116-18.

Laing, J.E. 1973. Evaluating the effectiveness of *Paracheyletia bakeri* (Acarina: Cheyletidae) as a predator of the two-spotted spider mite *Tetranychus urticae. Ann. Entomol. Soc. Am.* 66: 641-646.

Lal, B.S. and Rahman, M.F. 1975. Studies on the bionomics and control of the erinose mite, *Eriophyes litchii* Keifer (Acarina: Eriophyidae). *Pesticides* 9 (11): 49-54.

Lal, S.S. 1982. Influence of weather factors on the population of spider mite (Acari-Tetranychidae) and thrips (Thysanoptera) on cassava in Kerala. *Indian J. Acarol.* 7: 5-10.

Lall, B.S., Dutta, G.P. and Jaswal, A.P. 1965. Bionomics and control of red spider mite, *Tetranychus telarius* Linn. (Acarina: Tetranychidae) *Indian J.Ent.* 27: 389-392.

Lanciani, C.A. 1979. Water mite-induced mortality in a natural population of the mosquito *Anopheles crucians* (Diptera: Culicidae). *J. Med. Entomol.* 15: 529-532.

Lanciani, C.A. 1982. Parasite-induced reductions in the survival and reproduction of the backswimmer *Buenoa scimitra* (Hemiptera: Notonectidae). *Parasitol.* 85: 593-603.

Lanciani, C.A. 1986. Reduced survivorship in *Dasyhelea mutabilis* (Diptera: Ceratopogonidae) parasitized by the water mite *Tyrrellia circularis* (Acariformes: Limnesiidae). *J. Parasitol.* 72: 613-614.

Lange, A.B.; Dvozdovskii, E.M.; Bushkovskaya, L.M. 1974. Collecting and releasing *Anystis*. *Zaschita Rast.:* 33-34.

Langhe, A.B. and Natzkii, K.V. 1977. The mite *Varroa* and the methods of controlling it. In: *Varroasis-A Honeybee Disease* Apimondia Publishing House, Bucharest: 40-46.

Lawson, A.B. and Walde, S.J. 1993. Comparison of the responses of two predaceous mites, *Typhlodromus pyri* and *Zetzellia mali,* to variation in prey density. *Exp. App. Acarol.* 17: 811-821.

Lee, W.T., Ho, C.C. and Lo, K.C. 1990. Mass production of phytoseiids: I. Evaluation on eight host plants for mass rearing of *Tetranychus urticae* Koch and *T. kallzawai* Kishida (Acarina: Tetranychidae). *J. Agric. Res. China*. 39: 121-132.

Leigh, T.R., Maggi, V.L. and Wilson, LT. 1984. Development and use of a machine for recovery of arthropods from plant leaves. *J. Econ. EntomoI.* 77: 271-276.

Leite, R.C., Faccini, J.L.H. and Cosla, A.L. 1989. Avaliacao de uma tecnica in vivo para media a infestacao por acaros do genero Raillietia Trouessart (Acari) em bovinos. *Mem. Inst. Oswaldo Cruz* 84 (Suppl. IV): 309-311.

Leopold, R.A. 1998. Cold Storage of insects for integrated pest management. In: *Temperature sensitivity in insects and application in integrated pest management.* (eds. Hallman, G.J. and Denlinger, D.L.), Westview Press, Boulder: 235-268.

Lesna, I., Sabelis, M.W., Bolland, H.R. and Conijn, C.G.M. 1995. Candidate natural enemies for control of *Rhizoglyphus robini* Claparede (Acari: Astigmata) in lily bulbs: exploration in the field and pre-selection in the laboratory. *Exp. App. Acarol.* 19: 655-669.

Lester, P.J., Thistlewood, H.M.A. and Harmsen, R. 2000. Some effects of pre-release host-plant on the biological control of *Panonychus ulmi* by the predatory mite *Amblyseius fallacies Exp. Appl. Acarol.* 24(1) 19-33.

Lin, A. and Ridsdill-Smith, T.J. 2000. Cold storage of *Halotydeus destructor* (Acari: Penthaleidae) for use in experiments. *Exp. Appl. Acarol.* 24(2): 123-133.

Lin, H.; Otis, G.W.; Scott-Dupree, C.D. 1996. Comparative resistance in Buckfast and Canadian stocks of honey bees (*Apis mellifera* L.) to infestation honey bee tracheal mites (*Acarapris woodi* (Rennie). *Exp. Appl. Acarol.* 20: 87-101.

Lindquist, E.E. 1986. The world genera of Tarsonemidae (Acari: Heterostigmata): a morphological, phylogenetic and systematic revision, with a reclassification of family group taxa in the Heterostigmata. *Mem. Entomol. Soc. Canada.* 136: 1-517.

Linford, M.B. and Oliveira, J.M. 1938. Potential agents of biological control of plant parasitic nematodes. *Phytopathology* 28: 14.

Liu, T.S., Hung, Y.C., Twu, J.S. and Hsieh, J.H. 1998. Steam sterilization of soil for the control of bulb mites. *Pl. Prot. Bull.* 40: 241: 249.

Logan, J.A. 1982. Recent advances and new directions in phytoseiid population models. In: *Recent Advances in Knowledge of phytoseiidae* (ed Hoy, M.A.), University of California, Division of Agricultural Science, Publ. No. 3284: 49-71.

Lubinevski, Y., Stern, Y., Slabezki, Y, Lenski, Y., Ben-Yossef., H. and Gerson, U. 1988. Control of *Varroa jacobsoni* and *Tropilaelaps clareae* mites using Mavrik™ in *A. mellifera* colonies under subtropical and tropical climates. *Am. Bee J.* 128: 48-52.

Macfadyen, A. 1955. A comparison of methods for extracting soil arthopods. In: *Soil Zoology* (ed. Kevan, D.K.), Butterworths, London: 315-332.

Macfadyen, A. 1961. Improved funnel type extractors for soil arthopods. *J. Anim. Ecol.* 30: 171-184.

Macfadyen, A. 1962. Soil arthropod sampling. *Adv. Ecol. Res.* 1: 1-34.

MacPhee, A.W. and Sanford, K.H. 1954. The influence of spray prgorammes on the fauna of apple orchards in Nova Scotia VII. Effects on some beneficial arthropods. *Can Entomol.* 86: 128-135.

Mallea, A.R., Macola, G.S., Garcia Saez, J.G. and Lamate, S.J. 1983. Localizacion de *Pyemotes* sp. (Acarina: Pyemotidae) sobre *saissetra oleae* Bern. (Homoptera: Lecaniidae). *Intersectum* 15: 8-10.

Mallik, B. and Channabasavanna, G.P. 1983. *Indian J. Acarol.* 8: 1-12.

Mallik, B., Vaidya, H. and Harish Kumar, M. 1999: *J. Acarol.* 15: 15-17.

Manjunatha, M., Hanchenal, S.G. and Kulkarni, S.V. 1995. *Abs. V Nat. Sym. Acarol.* 20p.

Manjunatha, M., Hedge, M., Patil, B.V. and Lingappa, S. 1995. *Abst. V Nat. Symp. Acarol*, P. 72.

Mankau, R. 1980. Biological control of nematode pests by natural enemies. *Ann. Rev. Phytopathol.* 18: 415-440.

Markwick, N.P., Wearing, C.H. and Shaw, P.W. 1990. Pyrethroid insecticides for apple pest control. l. Development of pyrethroid resistant predatory mites. *Proc. NZ Weed Pest Control Conf.* 43: 296-300.

Martin, S.J. 1994. Ontogenesis of the mite *Varroa jacobsoni* Oud in worker brood of the honeybee *Apis mellifera* L. under natural conditions. *Exp. App. Acarol.* 18: 87-100.

Martin, S.J. 1995. Ontogenesis of the mite *Varroa jacobsoni* Oud. in drone brood of the honeybee *Apis mellifera* L. under natural conditions. *Exp. App. Acarol.* 19: 199-210.

Marwick, N.P. 1986. Detecting variability and selecting for pesticide resistance in two species of phytoseiid mites. *Entomophaga* 31: 225-236.

Mathur, R.B. and Mathur, S. 1982. Mites associated with stored grains products in Haryana, India. *Indian J. Acarol.* 7: 44-52.

Mathur, S. 1979. A new record of *Tyroborus lini* Oudemans form India. *Acarol. News Let.* 9: 3.

Matioli, A.L. 1997. Biologia de *Acarophenax lacunatus* (Cross and Krantz) (Prostigmata: Acarophennacidae) e Seu Potential de Parasitismo Sobre *Rhizopertha dominica* (F.) (Coleoptera: Bostrichidae). *Universidade Fedral de Vicosa,* Vicosa, Brazil.

Matheson, A. 1993. World bee health report. *Bee Wld.* 74: 176-212.

Matthysse, J.G., Van Vreden, F., Purnasari, A., Jones, C.J., Netherton, H.F. and McClain, O.S. 1975. Comparative susceptibility of the chorioptic mange mite, northern fowl mite and brown dog tick to acaricides. *Search Agric, Cornell Univ. Agric. Expt. Stn.* 14: 3-30.

Maunsell, K., Wraith, D.G. and Cunnington, A.M. 1968. Mites and house-dust allergy in branchial asthma. *Lancet* 1: 1267–1270.

Maurya, K.R., Jamil, Z. and Dev, B. 1983. Mite infestation pattern in various stored products in Lucknow (India). *Biol. Mem.* 7(2): 116-120.

McClure, H.E. 1989. Occurrence of feather mites (Proctophyllodidae) and 1ong birds of Ventura County Lowlands, California. *J. Field Ornithol.* 60: 431-450.

McGarry, J.W. and Trees, A.J. 1991. Trap perches to assess the activity of pyrethrins against the poultry realmite *Dermanyssus gallinae* in cage birds. *Exp. Appl. Acarol.* 12: 1-7.

McMurthy, J.A. and Seriven, G.T. 1966. Effect of artificial foods on reproduction and development of four species of phytoseiid mites. *Ann. Entomol. Soc. Am.* 59: 267-269.

McMurthy, J.A., Huffaker C.B. and Van de Vrie, M. 1970. Ecology of tetranychid mites and their natural enemies. A review. I. Tetranychid enemies their biological characters and the impact of spray practices. *Hilgardia* 40: 331-390.

McMurthy, J.A. 1982. The use of phytoseiids for biological control: progress and future prospects: In: *Recent Advances in knowledge of the Phytoseiidae* (ed. Hoy. M.A.), University of California, Division of Agricultural Sciences Publication no. 3284: 23-48.

McMurthy, J.A. and Rodriguez, J.G. 1987. Nutritional ecology of phytoseiid mites. In: *Nutritional Ecology of Insects, Mites and Spiders* (eds. Slansky, F. and Rodriguez, J.G.), Wiley, Chichester: 609-644.

Megevand, B.; Klay, A.; Guanvasson, D.; Paraiso, G. 1993. Maintenance and mass rearing of phytoseiid predators of the casseva green mite *Exp. App. Acarol.* 17: 115-128.

Mendel, Z. and Gerson, U. 1982. Is the mite *Lonyia Formosa* cooreman (Prostigmata: Tydeidae) a sanitizing agent in citrus groves ? *Oecol. Appl.* 3: 47-51.

Michael, A. D. 1881. Observations on the life histories of Gamasinae with a view to assist in their more exact classification. *J. Linn. Soc. (Zool.)* 15: 297-309.

Mihm, J.A. and Chiang, H.C. 1976. Laboratory studies on the life cycle and reproduction of some soil and manure-habiting predatory mites (Acarina: Laelapidae). *Pedobiologia* 16: 353-363.

Milani, N. 1995. The resistance of *Varroa jacobsoni* Oudemans to pyrethroids: a laboratory assay. *Apidologie* 26: 415-429.

Misra, K.K., Sarkar, P.K., Das, T.K. and Som Choudhary, A.K. 1990. Incidence of *Tetranychus cinnabarinus* (Boisd.) (Acari: Tetranychidae) on some selected accessions of brinjal with special reference to the physical basis of resistance. *Indian Agricult.* 34(3): 177-185.

Miyamoto, T., Oshima, S., Mizuno, K., Sasa, M. and Ishizaki, T. 1969. Cross antigenicity among six species of dust mites and house dust antigens. *J. Allergy* 44: 228-238.

Mohanasundaram, M. 2001. Phylogeny of Acarina and general classification. In: *Biodiversity and management of phytophagous mites* (eds. Rabindra R.J, Karuppuchamy, P., Umapathy, G., Chinniah, C., Balasubramani, V.), Deptt. of Agri. Entomol, TNAU, Coimbatore: 18-20.

Moore, D and Alexander, L. 1990. Resistance of coconuts in St. Lucia to attack by coconut mites, *Eriophyes guerreronis* Keifer. *Trop. Agric.* 67: 33-36.

Morewood, W.D. 1992. Cold storage of *Phytoseiulus persimilis* (Phytoseiidae). *Exp. Appl. Acarol.* 13: 231-236.

Morris, H.M. 1927. The insect and other invertebrate fauna of arable land at Rothamsted. Part II. *Ann. Appl. Biol.* 14: 442-464.

Morwick, N.P.; Wearing, C.H.; Shaw, P.W. 1990. Pyrethroids insecticides for apple pest control: 1. Development of pyrethroid resistant predatory mites. *Proc. N.Z. Weed Pest Control Conf.* 3: 296-300.

Moser, J.C., Smiley, R.L. and Otvos, I.S. 1987. A new *Pyemotes* (Acari: Pyemotidae) reared from the Douglas-fir cone moth. *Int. J. Acarol.*, 13, 141-147.

Mossadegh, M.S. and Komeili, B.A. 1986. *Euvarroa sinhai* Delfinado and Baker (Acarina: Mesostigmata): a parasitic mite on *Apis florea* F. in Iran. *Am. Bee J.* 126: 684-85.

Mourier H. and Poulsen, K.P. 2000. *J. Stored Prod. Res.* 36: 309-318.

Mozes-Koch, R., Slabezki, Y., Efrat, H., Kalev, H., Kamer, Y., Yakobson, B.A. and Dag, A. 2000. First detection in Israel of fluvalinate resistance in the varroa mite using bioassay and biochemical methods. *Exp. Appl. Acarol.* 24 (1): 35-43.

Mullen, C.R. 1975. Predation by water mites (Acarina: Hydrachnellae) on the immature stages of mosquitoes. *Mosq. News* 35: 168-171.

Mullen, C.R. 1977. Acarine parasites of mosquitoes. IV. Taxonomy, life history and behavior of *Thyas barbigera* and *Thyasides sphagnorum* (Hydrachnellae: Thyasidae). *J. Med. Entomol.* 13: 475-485.

Muma, M.H. 1975. Mites associated with citrus in Florida. *Univ. Fla Agric. Exp. St. Bull.* 640A.

Muraoka, M. and Ishibashi, N. 1976. Nematode feeding mites and their feeding behaviour. *Appl. Entomol.Zool.* 11: 1-7.

Murphy, P.W. 1958. The quantitative study of soil meiofauna. 1. The effect of sample treatment on extraction efficiency with a modified funnel extractor. *Entomologia Experimentalis et Applicata* 1: 94-108.

Murphy, P.W. 1962. *Progress in soil Zoology*, Butterworths, London.

Murphy, P. W. and Doncaster, C. C. 1961. In: *Research Methods in Soil. Zoology* (ed. Murphy, P. W.) London.

Murray, A.B. and Ferguson, A.C. 1983. Dust free bedrooms in the treatment of asthmatic children with housedust or house dust mite allergy: a controlled trial. *Pedialries* 71: 418-422.

Musk, A.W., Venables, K.M., Crook, B., Nunn, A.J., Hawkins, R., Crook, G.D.W., Graneck, B.J., Tee, R.D., Farrer, N., Johnson, D.A., Gordon, O.J., Darbyshire, J.H., Newman Taylor, A.J. 1989. Respiratory symptoms, lung function and sensitization to flour in a British bakery. *Br. J. Ind. Med.* 46: 636-642.

Nachman, G., Pederson, O.C., Munster-Swendsen, M. and Tonkiewicz, J. 1990. Sampling strategies for assessing the intra-plant density of the cassava green mite (*Mononychellus tanajoa*): A case study. *Exp. App. Acarol.* 9: 31-49.

Nachman, G., Skovgard, H., Tomkiewicz, J., Munster-Swendsen, M. and Pedersen, O.C.1993. Sampling strategies for assessing the overall density of the cassava green mite (*Mononychellus tanajoa*). *Exp. App. Acarol.* 17: 5-18.

Nageshchandra, B.K., Jagadish, P.S. Channabasavanna, G.P. and Nangia, N. 1999. *J. Acarol.* 15: 36-39.

Naitam, N.R., Patangrao, D.A. and Deshmukh, S.D. 1990. Resistance responses of chilli cultures to leaf curl. *PKV-Res. J.* 14(2): 206-207.

Nangia, N and Channabasavanna, G.P. 1983. *Abs. II All India Symp. Acarol*: 55.

Nangia, N and Channabasavanna, G.P. 1995, *J. Acarol.* 13: 45-48.

Nannelli, R., Liguori, M. and Castagnoli, M. 1983. Observazioni preliminari sulla biologia di *Euroglyphus (E.) maynei* (Coor.)

(Acarina: Pyroglyphidae) e sua distribuzione in ltalia. *Rediao* 66: 3-10.

Natarajan, K. 1988. Transport of yellow mite *Polyphagotarsonemus latus* by the cotton whitefly *Curr. Sci.* 57: 1142-1143.

Natuhara, Y. 1989. Housedust mites and allergy. *Jpn. J. Environ. Entomol. Zool.* 1: 40-54.

Nelson, E.E., Croft, B.A., Howett, A.J. and Jones, A.L. 1973. Toxicity of apple orchard pesticides to *Agistemus fleschneri. Environ. Entomol.* 2: 219-222.

Nene, Y.L. 1972. A survey of viral diseases of pulse crops in Uttar Pradesh. *Research Bulletin* No.4, G.B. Pant University of Agriculture and Technology, Pantnagar, Uttar Pradesh: 191 pp.

Newell, I.M. 1979. *Acarus tinctorius* Linnaeus 1767 (Trombidiidae).. In: *Recent Advances in Acarology* (ed. Rodriquez, J.G.) 2: 425-428.

Nottles, W.C., Morrison, R.K., Xie, Z., Ball, D., Shenkir, C.A. and Winson, S.B. 1982. Synergistic action of potassium chloride and magnesium sulfate on parasitoid wasp oviposition. *Science* 218: 164-166.

Nyein, M.M. and Zmarlicki, C. 1982. Control of mites in European bees in Burma. *Am. Bee J.* 122: 638-39.

Ochieng, R.S., Oloo, G.W. and Amboga, E.O. 1987. An artificial diet for rearing the phytoseiid mite, *Amblyseius teke* Pritchard and Baker. *Exp. App. Acarol.* 3: 169-173.

Oldfield, G.N., Reddy, M.V., Nene, Y.L. and Reed, W. 1981. Preliminary studies of the eriophyid vectors of mosaic. *Intl. Pigeonpea Newslet.* 1: 25-26.

Omar, H.I.H., Hegab, M.F. and Elsorady, A.E.M. 1993. The impact of intercropping cotton and cowpea on pest infestation. *Egyptian J. Agric. Res.* 71: 709-716.

Osaka, F., Kasuga, H., Sugita, M., Matsuki, H. and Miyake, T. 1985. A study on the relationship between mite IgE in serum and housing environment in school children. housing environment. *Jpn. J. Pub. Health* 32: 731-737.

Oshima S, Nakamura Y. and Sugita K. 1972. Population density of the mites of floor affected by the moisture content of the Tatami mat, with a particular reference for practical prevention. *Ann.. Rev Yokohana Inst. Health* 11: 62-69.

Oshima, S. 1967. Studies on the genus *Dermatophagoides* as floor mites, with special reference to the medical important. *Jpn..J. San. Zool.* 18: 213-215.

Overmeer, W.P.J. 1981. Notes on the breeding phytoseiid mites from orchards (Acarina: Phytoseiidae) in the laboratory. *Meded Fac. Landbouwwet Rijksuniv. Gent* 46: 503-509.

Overmeer, W.P.J. 1985. Rearing and handling. In: *Spider Mites: their biology, natural enemies and control* (eds. Helle, W. and Sabelis, M.W.), Elsevier, Amsterdam: 161-170.

Pande, YD. and Majummdar, M. 1990. *Abst. V Nat. Symp. Acorol.*: 18.

Pankiewicz-Nowicka, D., Boczek, J. and Davis, R. 1982. Some food preferences of *Acarus siro* L. *J. Georgia Ent. Soc.* 17(4): 491-495.

Parent, B. 1961. Effects de certains products antiparasitaires sun *Typhlodromus rhenanus* (Oudms) et *Mediolata mali* (Ewing), deux acariens predatenrs dr. Tetranyque rouge du pommier. *Ann. Entomol. Soc. Quebec.* 6: 55-61.

Parkinson, C.L., Barron, C.A., Barker, S.M., Thomas, A.C. and Armitage, O.M. 1991. Longevity and fecundity of *Acarus siro* on four field and eight storage fungi. *Exp. Appl. Acarol.* 11: 1-8.

Peace, D.M.C. 1983. Reproductive success of the mite *Acarus siro* L. on stored cheddar cheese of different ages. *J. Stored Prod. Res.* 19: 97-104.

Pemberton, R.W. and Turner, C.E. 1989. Occurrence of predatory and fungivorous mites in leaf domatia. *Ann. J. Bot.* 76:105-112.

Pena, J.E., Sharp, J.L. and Wysoki, M. 2002. *Tropical fruit pests and pollinators – Biology, economic importance, natural enemies and control*, CAB Internatural, U.K.

Peng,Y.S., Fang, Y., Xu, S. and Nsr M.E. 1987. Response of foster Asian honeybee (*Apis cerana* Fabr.) colonies of the brood of European honey bee *Apis mellifera* L.) infested with parasitic mite, *Varroa jacobsoni* Oudemans. *J. Invert. Pathol.* 49: 259-264.

Penman, D.R., Chapman, R.B. and Jessan, K.E. 1981. Effects of fenvalerate and azinphosmethyl on two spotted spider mite and phytoseiid mites. *Entomol. Exp. Appl.* 30: 91-97.

Pepys, J., Moira C. and Hargreave, F.E. 1968. Mites and hosue-dust allergy. *Lancet* 1: 1270-1272.

Perez, T.M. and Atyeo, W.T. 1984. Site selection of the feather and quill mites of Mexican parrots. In: *Acarology VI* (eds. Griffiths, O.A. and Bowman, C.E.), Ellis Horwood Ltd., Chichester: 563-570.

Perkins, B.D. 1973. Preliminary studies of a strain of the waterhyacinth mite from Argentina. In: *Proc. 2nd Int. Symp. Bioi. Cant. Weeds* (ed. R.H. Dunn), Commonwealth Agricultural Bureaux.

Peterson, P.C. 1975. An analysis of host parasite associations among feather mites (Acari-Analgoidae). *Misc. Publ. Entomb. Soc. Am.* 9: 237-242.

Petrushov, A.Z. 1991. Selection and mechanisms of ambush resistance in *Metaseiulus occidentalis* N. (Acarina: Phytoseiidae). *Modern Acarol.* 2: 741-748.

Petrushov AZ. 1992. Pyrethroid resistance in the predacious mite *Amblyseius barkeri*. *EPPO Bull.* 22: 471-473.

Pillai, R.R.P. 1957. Pests of stored fish and prawns. *Bull Central Res. Inst. Univ. Kerala. Trivendrum Series (National Sci.)* 5(3): 1-79.

Platts-Mills, T.A.E. and Chapman, M.D. 1987. Dust mites: immunology, allergic disease and environmental control *J. Allergy Clin Immunol.* 80: 755-775.

Platts-Mills, T.A.E. and DeWeek, A.L. 1989. Dust mite allergens and asthma: a worldwide problem (In. Workshop Report). *J. Allergy Clin. Immunol.* 83: 416-427.

Platts-Mills, T.A.E, Tovery, E.R., Mitchell, E.B., Moszoro, H., Nock. P. and Wilkins, S.R. 1982. Reduction of bronchial hyperactivity during prolonged allergen avoidance. *Lancet* 2: 675-678.

Poiani, A. 1992. Ectoparasitism as a possible cost of social life: a comparative analysis using Australian passerines (Passeriformes). *Oecologia* 92: 429-441.

Poinar, G. and Poinar, R. 1998. Parasites and Pathogens of Mites. *Ann. Rev. Entomol.* 43: 449-69.

Potts, M.E. and Rodriquez, J.G. 1979. *Rec. Adv. A carol.* 1: 201-210.

Poulin, R. 1991. Group living and infestation by ectoparasites in passerines. *Condor* 93: 418-423.

Prasad, V. 1974. *A catalogue of mites of India.* Indira Acarology Publishing house, Ludhiana 320pp.

Prasad, V.G. and Singh, R.K. 1981. Prevalence and control of litchi mite, *Aceria litchi* Keifer in Bihar. *Indian J. Ent.* 43: 67-75.

Pree, D.J. and Wagner, H.W. 1987. Occurrence of cyhexalin and dicofol resistance in the European red mite, *Panonychus ulmi* (Koch) (Acari: Tetranychidae), in Southern Ontario. *Can. Entomol.* 119: 287-290.

Proctor, H.C. 1997. Mating behavior of *Physolimnesia australis* (Halik) (Acari: Limnesiidae), a non-parasitic, rotifer-eating water mite from *Australia J. Arachnol.* 25: 321-325.

Proctor, H.C. and Pritchard, G. 1990. Prey detection by the water mite *Unionicola crassipes* (Acari: Unionicolidae). *Fresh water Biol.* 23: 271-279.

Pulpan, J. and Verner, P.H. 1965. Control of tyroglyphid mites in stored grain by the predatory mite, *Cheyletus eruditus* (Schrank). *Can.J. Zool.* 43: 417-432.

Purvis, M.J., Collier, D.C. and Walls, D. 1964. *Laboratory techniques in botany.* Butterworth and Co. Ltd., London: 371 pp.

Putatunda, B.N. 1979. *Mites associated with some important birds in West Bengal (India).* Ph D. Thesis. B.H.U., Varanasi pp 1-164 + I-XVIII.

Putatunda, B.N. 2001. Mites (Acarina) in post harvest storage of some food grains in Himachal Pradesh, India. *Pl. Protec. Bull.* 54(3&4).

Putatunda, B.N. Gupta, S.K. and Singh, J. 1981. Mites associated with poultry in West Bengal. In: *Contributions to Acarology in India* Acarol. Soc. India. Bangalore 85-93.

Putatunda, B.N., Mathur, S. and Mathur, R.B. 1997. Mites (Acarina) associated with stored products in Haryana, India – Taxonomic study. In: *Recent Advances in Biosciences and Ocenography.* (Eds. Aggarwal, V.P. *et al.*), Soc. Biosciences India: 213-216.

Putatunda, B.N.; Gupta, S.K. and Singh, J. 1989. Acarine associates of birds in West Bengal, India. *Progress in Acarology* 2: 309-314.

Putman, W.L. 1962. Life history and behaviour of predacious mite *Typhlodromus (T.) candiglans* Schuster (Acarina: Phytoseiidae) in Ontario, with notes on the prey of related species. *Can. Entomol.* 94: 163–177.

Puttaswamy and Channabasavanna, G.P. 1979. Effect of temperature and relative humidity on the development and oviposition of *Tetranychus ludeni. Indian J. Acarol.* 4(1): 31-39.

Puttaswamy. 1978. Ph. D. Thesis. UAS, Bangalore.

Pyke, G.H., Pulliam, H.R. and Charnov, E.L. 1977. Optimal foraging: a selective review of theory and tests. *Q. Rev. Biol.* 52: 137–154.

Rabindra, R.J., Karuppuchamy, P., Umapathy, G., Chinniah, C. and Balasubramani, V. 2001. *Biodiversity and Management of Phytophagous Mites*, Dept. of Agri. Entomology, TNAU, Coimbatore: 310 pp.

Rai, S.N. and Singh, J. 1999. *J. Acarol.* 14: 4-7.

Ramaraju, K., Velusomy, R. and Sathakathulla, S.S. 1997. Genetic evaluation of red spider mite resistance in *Abelmoschus* sp. *TNAU Newsletter* 27(2): 1.

Ranganath, H.R. and Channabasavanna, G.P. 1988. House dust mites from Bangalore, India–A quantitative analysis. In: *Mite Allergy–A world wide problem* (eds. De Week, A.L. and Todt, A.), WCB Institute of Allergy, Brussel: 21-22.

Rangel AF. 1901. Cuarto informe acerca del picudo del algodon (*Insanthonomus grandis*, l.C.C.U.). *Biol. Comis. Parasitol. Agri.* 1: 245-261.

Ranjana, R. 2002. *Studies on the nematophogy in acarine fauna and their feeding potential on seed gall nematode*. M.Sc. Thesis, CCSHAU, Hisar. 71+xiv pp.

Ranjana, R., Gulati, R. and Walia, K.K. 2003. Influence of prey size, prey density and test area on predatory potential of two nematophagous mites. In: *5th Nat. Sym. Biocontrol Agents for Sustainable Management of Pests*, Pantnagar, Uttaranchal.

Ranjana, R., Gulati, R. and Walia, K.K. 2005. Feeding potential of *Tyrophagus putrescentiae* Schrank (Acarina: Acaridae) on seedgall nematode, *Anguina tritici* (Steinbuch) Chitwood. *J. Insect Sci.* 19: 32-35.

Rao, P.V.R., Narsimharao, P.L. and Tirumalrao, 1957. Association of pinkish brown blotch of citrus fruits with eriophyid (*Phyllocoptruta oleivorus* (Ashmead). *Indian J. Ent.* 19: 146-147.

Rao, V.P. and Sankaran, J. 1969. The scale insects of sufarcane. In: *Pests of sugarcane* (eds. Williams, J.R., Metcalfe, J.R., Mungomery, R.W. and Mathes, R.), Elsevier, Amsterdam: 325-342.

Rasmy, A. H, Abdel-Rahman, H.A. and Hussein, H.E. 1991. Suitability of different mite prey for the development of the predatory mite, *Phytoseiulus persimilis*. *Exp. Appl. Acarol.* 11: 89-91.

Rasmy, A.H., Elbagoury, M.E. and Reda, A.S. 1987. A new diet for reproduction of two predaceous mites *Amblyseius gossipi* and *Agistemus exsertus* (Acarina: Pyetosiidae, Stigmaeidae). *Entomophaga* 32: 277-280.

Rath, W., Delfindo-Baker, M. and Drescher, W. 1991. Observations on the mating behaviour, sex ratio, phoresy and dispersal of *Tropilaelaps clareae* (Acari: Laelapidae). *Intl. J. Acarol.* 17 (3): 201-208.

Rathi, A. 1991. *M.Sc. Thesis,* CCS Haryana Agricultural University, Hisar, India.

Rather, A.Q. 1999. *J. Acarol.* 15: 18-24.

Readshaw, J.L. 1971. An ecological approach to the control of mites in Austra1ian orchards. *J. Aust. Inst. Agric. Sci.* 41: 213-214.

Reddy, M.V., Raju, T.N. and Lenna, J.M. 1998. Diseases of pigeonpea. In: *The pathology of food and pasture legumes* (eds. Allen, D.J. and Lenne, J.M.) CAB International ICRISAT: 517-558 pp.

Reed, D.K., Burditt, A.K. and Crittenden, C.R. 1964. Laboratory methods for rearing rust mites (*Phyllocoptruta oleivora* and *Aculus pelekassi*) on citrus. *J. Ecol. Entomol.* 57: 130-133.

Reed, D.K. 1981. Control of mites by non-occluded viruses. In: *Microbial control of pests and plant diseases.* (ed. Burges, H.D.), Academic Press, New York: 427-432.

Reisen WK, Mullen GR. 1978. Ecological observations on acarine associates (Acari) of Pakistan mosquitoes (Diptera: Culicidae). *Environ. Entomol.* 7: 769-776.

Revsbech, P. and Andersen, G. 1987. Storage mite allergy among grain elevator workers. *Allergy* 42: 423-429.

Revsbech, P. and Ducholm, M. 1990. Storage mite allergy among bakers. *Allergy* 45: 204-208.

Riedl, H. and Shearer, P.W. 1991. Double leaf-disk residue assay for assessing the toxicity of repellent acarieides to spider mites (Acari: Tetranychidae). *Exp. Appl. Acarol.*: 149-157.

Riley, C.V. 1874. Description of two new subterranean mites. *Trans. Acad. Sci. St. Louis* 25:215-216.

Rishi, N.D. 1990. *Abst. V. Nat. Symp. Acarology*: 32-33.

Ritter W. 1981. *Varroa* disease of the honeybee *Apis mellifera*. *Bee Wld* 62: 141-153.

Robaux P. 1974. Recherches sur Ie développement et la biologie des acariens "Thrombidii dae" *Mem. Mus. Hist. Nat. S Aer. A. Zool* 85: 1-186.

Rockett, C.L. 1980. Nematode predation by oribated mites (Acari: Oribatida). *Int. J. Acarol.* 6: 219-224.

Rockett, C.L. and Woodring, J.P. 1966. Oribatid mites as predators of soil nematodes. *Ann. Ent. Soc. Am.* 59: 669-671.

Rodinovsky, S. and Krantz, G.W. 1961. The biology and ecology of granary mites of the pacific northwest. II. Techniques for laboratory observations and rearing. *Ann. Ent. Soc. Am.* 51: 512-518.

Rodriquez, J.G. 1978. *Recent Advances in Acarology*. (Vol I and II). Publ. By Academic Press, Inc., UK. 569 pp.

Rodriguez, J.G., Potts, M.F. and Patterson, C.G. 1984. Mycotoxin-producing fungi: effects on stored product mites. In: *Acarology VI*. (eds. Griffiths, D.A. and Bowman, CE.). Ellis Horwood, Chichester 1: 343-350.

Rodriguez, J.G., Wade, C.F. and Wells, C.N. 1962. Nematodes as a natural food for *Macrocheles muscaedomesticae* (Acarina: Macrochelidae) a predator of the housefly eggs. *Ann. Ent. Soc. Ani.* 55: 507-511.

Rodriquez, J.G., Singh, P. and Taylor, B. 1970. Manure mites and their role in fly control. *J. Med. Entomol.* 7: 335-341.

Rohde, C. J. 1956. A modification of the plaster-charcoal technique for the rearing. of mites and other small arthropods. *Ecology,* 37: 843-844.

Rosen, D. and Huffaker, C.B. 1983. An overview of desired attributes of effective biological control agents, with particular emphasis on mites. In: *Biological control of pests by mites* (eds. Hoy, M.A. *et al.*), University of California special publication no. 3304: 2-11.

Rosenthal, S.S. 1983. Current status and potential for biological control of field bindweed, *Convolvus arvensis* with *Aceria convolvuli.* In: *Biological control of pests by mites* (eds Hoy, M.A., Cunningham, G.L. and Knutson, L.):57-60.

Roush, R.T. and Hoy, M.A. 1978. Relative toxicity of permethrin to a predator, *Metaseiulus occidentalis* and its prey, *Tetranychus urticae. Environ. Entomol.* 7: 287-288.

Ruttner, V. 1983. Varroatosis in honeybees: Extent of infestation and effects. In: *Varroa jacobsoni Oud. Affecting honeybees: Present status and needs* (ed. Cavalloro, R.): 7-14.

Sabelis, M.W. 1981. Biological control of two spotted spider mites using phytoseiid predators. Part 1. Modeling the predator-prey interaction at the individual level. *Agricul. Res. Rpt. 1910 Wageningen,* Netherlands: 242pp.

Sabelis, M.W. 1985. Capacity for population increase. In: *Spider Mites, their biology, natural enemies and control* (eds. Helle, W. and Sabelis, M.W.), Elsevier, Amsterdam 1 B: 35-41.

Sabelis, M.W. 1990. How to analyze prey preference when prey density varies: A New method to discriminate between effects of gut fullness and prey type composition. *Oecologia* 82: 289-298.

Sabelis, M.W. and Bakker, F.M. 1992. How predatory mites cope with the web of their tetranychid prey: a functional view on dorsal chaetotaxy in the Phytoseiidae. *Exp. Appl. Acarol.* 16: 203-225.

Sabelis, M.W. and Dicke, M. 1985. Long range dispersal and searching behaviour. In: *Spider Mites, their biology, natural enemies and control* (eds. Helle, W. and Sabelis, M.W.), Elsevier, Amsterdam 1 B: 141-160.

Sabelis, M.W. and Bann, Van de. 1983. Location of distant spider mite colonies by phytoseiid predators: demonstration of specific kairornones emitted by *Tetranychus urticae* and *Panonychus ulmi. Entomol. Exp. Appl.* 33: 303-314.

Sadana GL. 1985. *Plant feeding mites in India.* Kalyani Publishers, New Delhi. 99 pp.

Salem, F.M. and Abo-Korah, S.M. 1981. Inter relationships between plant parasitic nematodes and soil acari IV. Interaction between *Meloidogyne javanica* and certain acarid mite species in relation to horse bean nematode infection. *Menonfeia. Agric. Res.* 4: 345-353.

Salt, G. and Hollick, F.S.J. 1944. Studies of wireworm populations. A consus of wireworms in pasture. *Ann. Appl. Biol.* 31: 52-64.

Sammataro, D. 1997. Studies on the control, behaviour and molecular markers of the fracheal mite (*Acaropris woodi* (Rennie) of honey bees (Hymenoptera: Apidae). Ph.D. Diss. Ohio State Univ., Columbus, O.H. *Ann. Rev. Ent.* 45: 125pp.

Sammataro, D., Gerson, U. and Needhans, G.2000. Parasitic mites of honey bees: Life History, Implications and Impact *Ann. Rev. Entomol.* 45: 519-548.

Samsoe-Petersen L, Bigler F, Bogenschutz H, Brun J. 1989. Laboratory rearing techniques for 16 beneficial arthopod species and their prey/host. *Zeitschr. pflanzenkrankheiten u. Pflanzenschutz* 96: 289-316.

Santos, M.A. 1976a. Evaluation of *Zetzellia mali* as a predator of *Panonychus ulmi* and *Aculus schlechtendali. Environ.Entomol.* 5: 187-191.

Santos, M.A. 1976b. Prey selectivity and switching response of *Zetzellia mali. Ecology* 57: 390-394.

Santos, M.A. 1991. Searching behaviour and associational response of *Zetzellia mali* (Acarina: Stigmacidae). *Exp. App. Acarol.* 11: 81–87.

Santos, P.F. and Whitford, W.G. 1981. The effects of microarthropods on litter decomposition in a Chihuahuan desert ecosystem. *Ecology*, 62: 654-663.

Santos, P.F., Phillips, J. and Whitford, W.G. 1981. The role of mites and nematodes in early stages of buried litter decomposition in a desert. *Ecology,* 62: 664-669.

Schamber, J.F. 1910. Grain itch (acaro-dermititis, urticaroides). A study of a new disease in this country. *J. Cutan. Dis.* 28: 67.

Schruft, G. 1971. *Haleupalus oliveri* nov. spec., eine Dornpalpenmilbe an Reben (*Vitis* spec.) (Acari: Cunaxidae). *Dtsch. Entomol. Z.* 18: 377-382.

Schutz, A.E. 1984. Reproduction and population dynamics of the parasitic mite *Varroa jacobsoni* Oud. and its dependence on the brood cycle of its host, *Apis mellifera* L. *Apidologie* 15: 401-420.

Schuster, R. and Murphy, P.W. 1991. *The Acari: Reproduction, Development and Life history stratergies.* Chapman Hall, London. 554 pp.

Scriven, G.T. and McMurtry, J.A. 1971. Quantitative productions and processing of tetranychid mites for large scale testing of predator production. *J. Econ. Entomol.* 64: 1255-1257.

Searls, E.M. 1928. A simple method for life history studies of root-feeding arthropods. *J. Agric. Res.* 36: 639-645.

Seeley, T.D. and Morse, R.A. 1976. The nest of the honeybee (*Apis mellifera*). *Insect Soc.* 23: 495-512.

Seeman, O.D. and Walter, D.E. 1995. Life history of *Afrocypholaclaps africana* (Evans) (Acari: Ameroseiidae), a mite inhabiting mangove flowers and phoratic on honeybees. *J. Aust. Entomol. Soc.* 34: 45-50.

Segur, J.B. 1953. Physical properties of glycerol and its solutions. In: *Glycerol* (eds. Miner, C.S. and Dalton, N.N.), Reinhold, New York: 269.

Sell, P. 1988. *Caloglyphus* sp. (Acarina: Acaridae), an effective nematophagous mite on root-knot nematodes (*Meloidogyne* spp.). *Nematologica* 34: 246.

Sengbusch, H. G. 1954. Studies on the life history of three Oribatid mites with observations on other species. *Ann. Ent. Soc. Am.* 47: 646-667.

Seth, M.L. 1962. Transmission of pigeonpea sterility by a eriophyid mite. *Indian Phytopathol.* 15: 225-227.

Severin, H.C. 1944. The grasshopper mite, *Euthrombidium trigonum* (Hermann) an important enemy of grasshoppers. *South Dakota Agric. Exp. St. Techn. Bull.* 3: 1-36.

Sevilla, V.J. 1963. Observations on the life history and habits of three new acarine pests of honey bees in the Philippines. *University of Phillippines, B.Sc. thesis.*

Shah, F.A. and Shah, T.A. 1988. *Tropilaelaps clareae* a serious pest of honey bees; flour dusting control for varroa disease in Kashmir. *Bee Wld* 72 (4): 161-164.

Sharma, R.D. 1971. Studies on the plant parasitic nematode *Tylenchorhnchus dubius. Meded. Landbouww.* Wageningen 71: 1-154.

Sharma, A. and Kushwaha, K. S. 1983. *Abst. All India Symp. Acarol*: 19.

Sharma, A. and Srinivasa, N. 2004. *Brown wheat mite Petrobia latens (Muller) (Acari: Tetranychidae)*, AINP (Agricultural Acarology), UAS, Bangalore: 12 pp.

Sharma, N.K. and Sadana, G.L. 1983. *Abst. /I All India Symp. Acarology*: 54.

Sharma, N.K. and Sadana, G.L. 1989. *J. Acarol.* 11: 51-54.

Sharma, O.P.; Garg, R. and Dogra, G.S. 1983. Efficacy of formic acid against *Acarapis woodi* (Rennie). *Indian Bee J.* 45(1): 1-2.

Sharma, R D. 1971. Studies on the plant parasitic nematode *Ty1enchorhynchus dubius. Meded. Landbouwhogeschoo1, Wageningen.* 71: 98-104.

Sharov, A.G. 1966. *Basic arthropodan stock with special reference to insects*. Pergamon Press, Oxford: 271 pp.

Shastri, U.V. and Deshpande, P.D. 1983. Ear mites *Raillietia manforedi* Domrow, 1980. Sic and *Psosoptes cuniculi*.(Railliet. 1893) Sweetman 1958 Sic] in goats in Marathwada (Maharashtra) India. *Trop. Vet. Anim. Sci. Res.*1 (1): 94-96.

Shimer, H. 1868. Notes on the "apple bark louse" (*Lepidosaphes conchiformes*, Gmelin sp.) with a description of a supposed new *Acarus. Trans. Ann. Entomol. Soc.* 1: 361-374.

Singer, G. 1964. A simple aspirator for collecting small arthropods directly into alcohol. *Ann. Ent. Soc. Am.* 57: 796-798.

Singh,A. and Putatunda, B.N. 1999. Mites associated with poultry bed in Hisar (Haryana). *J. Acarol.* 15: 73-76.

Singh, M. 1985. *Biology and feeding behavior of an acarid mite, Suidasia nesbitti (Hughes).* M.Sc. Thesis, CCSHAU, Hisar: pp.

Singh, M. 1990. *Qualitative and quantitative changes in some food grains due to mite infestation.* Ph.D. Thesis, CCSHAU, Hisar: 94+xiii pp.

Singh, M. and Gulati, R. 1998. A report on the seasonal incidence of house dust mites in Haryana, India. *Ann. Agri–Bio Res.* 3 (2): 231-233.

Singh, M. and Gulati, R. 2001. Minimization of losses due to mites in major pulses during storage. *Final Technical Report.* CSIR, New Delhi: 1-129.

Singh, P. 1988. *Bionomics of Aceria litchii (Keifer) (Prostigmata. Eriophyidae).* Ph.D. Thesis. Bidhan Chand Krishi Vishwavidhalya. Kalyani: 185.

Sinha, R.N. 1964. Mites of stored grains in Western Canada. Ecology and Survey. *Proc. Entomol. Soc. Manitoba* 20: 19-33.

Sircom J. 2000. Photographic sampling: a photographic sampling method for mites on plants. *Exp. App. Acarol.*: 55-61.

Skovgard, H., Nachman, G., Tomkiewicz, J. and Munster-Swendsen, M. 1993. The dynamics of the cassava green mite *Mononychellus tanajoa* in Kenya–A field study. *Exp. App. Acarol.* 17: 59-76.

Smith, B.P. 1983. The potential of mites as biological control agents of mosquitoes. In: *Biological Control of Pests by Mites* (eds. Hoy, M.A., Cunningham, C.L. and Knutson, L.), University of California Special Publication no. 3304: 79-85.

Smith, B.P. 1988. Host parasite interaction and impact of larval water mites on insects. *Ann. Rev. Entomol.* 33: 487-505.

Smith, B.P. 1990. Hydrachnida. In: *Freshwater macroinvertebrates of northestern North America* (eds. Peckarsky, B.L. *et al.*), Cornell University Press, Ithaca: 290-334.

Smith, B.P. and McIver, S.B. (1984) The impact of *Arrenurus danbyensis* (Mullen) (Acari: Prostigmata; Arrenuridae) on a population of *Coquillettidia perturbans* (Walker) (Diptera: Culicidae). *Can. J. Zool.* 62: 1121-1134.

Smith, I.M. and Cook, D.R. 1991. Water mites. In: *Ecology classification of North American fresh water invertebrates.* (eds. Thorp, J.H. and Covich, A.P.), Academic Press, Inc., New York, 523-592.

Smith, I.M. and Oliver, D.R. 1986. Review of parasitic associations of larval water mites (Acari: Parasitengona: Hydrachnida) with insect hosts. *Can. Entomol.* 118: 407-72.

Solomon, M.E. 1946. Tyroglyphid mites in stored products, Nature and amount of damage to wheat. *Ann. Appl. Biol.* 33: 280-289.

Solomon, M.E. 1969. Experiments on predatory prey interactions of storage mites. *Acarologia.* 11: 484-503.

Solomon, M.E. 1972. Tyroglyphid mites in stored products: Methods for the study of population density. *Ann. Appl. Biol.* 69: 71-77.

Solomon, M.E. and Cunnington, A.M. 1964. Rearing acarid mites. *Acarologia* 6: 399-403.

Somchaudhary, A.K. and Mukherjee, A.B. 1971a. Bionomics of *Acaropsis docta* (Berlese) (Acarina: Cheyletidae), a predator on the eggs of some pests of stored grains. *Indian J. Ent.* 33 (3): 79-81.

Somchaudhary, A.K. and Mukherjee, A.B. 1971b. Observations on the seasonal abundance of some common stored grain mites in India. *Indian J. Ent.* 33: 350-357.

Sorensen, J.T., Kinn, D.N. and Doutt, R.L. 1983. Biological observations on *Bdella longicornis*: A predatory mite in California Vineyards (Acari: Bdellidae). *Entomography* 2:297-305.

Sorrensen, J.J., Doutt, R.L. and Cate, J.R. 1976. Biology of the mite *Anystis agilis* (Acarina: Anystidae): a California Vineyards predators. *Ann. Entomol. Soc. Ann.* 69: 905-910.

Southwick, E.E.; 1988. Menthol. *Am. Bee J.* 128(5): 369-370.

Southwood, T.R.E. 1968. *Ecological methods with particular reference to the study of insect populations.* Chapmen and Hall, New York.

Spieksma, F.T. M. 1990. Mite biology. *Clin. Rev. Allergy* 8: 31–48.

Spieksma, F.T.M. 1967. Occurrence and properties of the house-dust mite, *Dermatophagoides pteronyssinus. Int. Rhin.* 5: 162-167.

Spieksma, F.T.M. and Voorhorst, R. 1969. Comparison of skin reactions to extracts of house dust mites and human skin scales. *Acta Allergol.* 24: 124-146.

Spiewal, R., Gora, A. and Dutkiewicz, J. 2001. *Ann. Agric. Environ. A1ed.* 8(1): 51-56.

Springer, M. and Gerecke, R. 1992. The male of Neotorrenticola plumipes Lundblad, 1953 (Acari, Actinedida, Limnesiidae) from Costa Rica. *Studies on Neotropical Fauna and Environment* 27 (4): 243 – 252.

Srinivasa, N., Mallik, B. and Gowda, C.C. 2004. *Aceria cajani* (Acari: Eriophyidae) *Transmitted Pigeonpea sterility Mosaic Disease*, AINP (Agricultural Acarology), UAS, Bangalore: 24pp.

Stechmann, D.H. 1978. Eiablage, Parasitismus and post parasitische Entwicklung von *Arrenurus* – Arten (Hydrachnellae, Acari), *Z. Parasitenkd* 57: 169-188.

Steinhaus, F.A. 1959. Possible virus disease in European red mite. *J. Insect Pathol.* 1: 435-37.

Stemkraus, D.C. and Cross, E.A. 1993. Description and life history of *Acarophenax mahunkai* n. sr. (Acari, Tersonemina: Acarophenaciedae), an egg parasite of the leser meal warm (Coleoptera: Tenebrionidae). *Ann. Entomol. Soc. Am.* 86: 239-224.

Stephen, W.A. 1968. Mites a beekeeping problem in Vietnam and India. *Bee Wld.* 49:113-120.

Sternlicht, M. 1966. Trials in the control of the citrus bud mite, *Aceria sheldoni* Ewing. *Israel J. Agr. Res.* 16(3): 115-124.

Stoner, AK, Frank, JA and Gentile, GA. 1968. The relationship of glandular hairs on tomatoes spider mite resistance. *Proc. Am. Soc. Hort. Sci.* 93: 532-538.

Storm, V. L., WS-Bien, Z. and Varenkaml, H. 1924. Experimentalle allergische kranksheiteu (asthma bronchiale, rhinitus vasomotoria). *Z. Immunf. Exp. Ther.* 40: 552-569.

Strickler, K. and Croft, B.A. 1982. Selection for permethrin resistance in the predatory mite *Amblyseius fallacis*. *Entomol. Exp. App.* 31: 339-345.

Strong, W.E. and Croft, B.A. 1993. Predaceous phytoseiid mites associated with spider mites on hops in the Willamette Valley, Oregon. *J. Ent. Soc. Brit. Col.*

Strong, R.G., Pieper, G.R. and Sbur, D.E. 1959. Control and prevention of mites in granvy and Vice weevil cultures. *J. Econ. Entomol.* 52: 443-446.

Sturhan, D. and Hampel, G. 1977. (Plant parasitic nematodes as prey of the bulb mite, *Rhizoglyphus echinopus* (Acarina: Tyroglyphidae)]. *Anzieger fur schadlingskunde, Pflanzen-schutz umweltschutz* 50: 115-118.

Swan D.C. 1934. The hay itch mite..*J. Agric. Soc. Aust.* 37:.1280-1299.

Szlendak, E., Conyers, C., Muggleton, J. and Thind, B.B. 2000. Pirimiphos methyl resistance in two stored product mites, *Acarus siro* and *Acarus farris*, as detected by impregnated paper bioassay and esterase activity assays. *Exp. Appl. Acarol.* 24(1): 45-54.

Taha, H.A., El-Naggar, M.E.E., Abou-El-Nagaga, M.M. and Soliman, S.M. 1988. Effect of different prey species on the development and fecundity of the predacious mite, *Neocunaxoides andrei* Baber and Hoff (Acari: Cunaxidae). *Agri.Res Rev.*66: 129-135.

Takafuji, A. and Chant, D.A. 1976. Comparative studies of two species of predacious phytoseiid mites (Acarina: Phytoseiidae), with special reference to their responses to the density of their prey. *Res. Pop. Ecol.* 17: 255-310.

Tamura, H. and Enda, N. 1980a. Life histories of three species of nematode feeding mesostigmatid mites associated with the pine sawyer beetle *Monochamus alternatus. J. Jpn. For. Soc.* 62: 301-307.

Tamura, H. and Enda, N. 1980b. Mesostigmatid mites associated with Japanese pine sawyer beetle. *Jpn. J. Appl. Ent. Zool.* 24: 54.

Tandon, P.L. and Lal, B. 1976. New record of the predatory mites on mango mealy bug, *Drosicha mangiferae* Green (Margarodidae: Hemiptera). *Current Sci.* 45: 566-567.

Tanigoshi, L.K. 1982. Advances in the knowledge of the biology of the phytoseiidae. In: *Recent advances in knowledge of Phytoseiidae* (ed. Hoy, M.A.). Univ. Calif. Berkeley Publ. No. 3284: 1-22.

Tanigoshi, L.K. and Davis, R.W. 1978. An ultrastructural study of *Tetranychus medanidli* feeding injury to the leaves of Red Delicious apple (Acarina: Tetranychidae). *Int. J. Acarol.* 4: 47-56.

Tanigoshi, L.K., Nishio-wong, J.Y. and Fargerlund, J. 1984. *Euseius hibisci*: Its control of citrus thrips in Southern California citrus orchards. In: *Acarology VI* (eds. Griffiths, D.A. and Bowman, C.E.), Ellis H orwood, Chichester 2: 699-702.

Taylor, R.N. 1975. *Contribution to the biology and ecology of house dust mites*. Doctoral Thesis Univ. Glasgow: 98.

Teho, E.O., Husman, K., Vohlonen, I., Rautalahti, M. and Tukiainen, H. 1985. Allergy to storage mites or cow dander as a cause of rhinilis among Finnish dairy farmers. *Allergy*. 40: 23-26.

Tewarson, N.C., Singh, A. and Engels, W. 1992. Reproduction of *Varroa jacobsoni* in colonies of *Apis cerana indica* under natural and experimental conditions. *Apidologie* 23: 161-171.

Thind, B.B. and Muggleton, J. 1998. A new bioassay method for the detection of resistance to pesticides in the stored product mite, *Acarus siro* (Acari: Acaridae) *Exp. Appl. Acarol.* 22: 543: 552.

Thirugnanasunthaun, K. and Amarasinghe, L.D. 1990. Clonal susceptibility and population dynamics of tea red spider mite, *Oligonychus coffeae* Nictor (Acarina: Tetranychidae) under laboratory conditions. *Shrilank J. Tea Sci.* 59: 9-15.

Thistlewood, H.M.A. 1991. A survey of predatory mites in Ontario apple orchards with diverse pesticide programs. *Can. Entomol.* 123: 1163–1174.

Thistlewood, H.M.A., Pree, O.J. and Crawford, L.A. 1992. Comparison of slide dip and petridish assays for measuring resistance to permethrin in *Amblyseius fallacis* (Acari: Phytoseiidae). *J. Econ. Entomol.* 85: 2051-2057.

Thwaite, W.G. 1991. Resistance to clofentezine and hexythiazox in *Panonychus ulmi* from apples in Australia. *Exp. Appl. Acarol.* 11: 73-80.

Tomczyk, A. and Kropczynska, K. 1985. Effects on the host plant. In: *Spider mites, their biology, natural enemies and control* (eds. Halle, W. and Sabelis, M.W.), Elsevier, Amsterdam: 317-329.

Tseng, Y.H. 1984. Mite associated with weeds, paddy rice and upland rice fields in Taiwan. In: *Acrology VI* (eds Griffiths, D.A. and Bowman, C.E.), Ellis Horward, Chickester. 2: 770-780.

Tullgren, A. 1918. Ein sehr ein facher Ausks eapparat fur terricole Tierformen. *Z. angew. Ent.* 4: 149-150.

Uchikoshi, S., Kimura, H., Nomura, K., Chien, C., lida, M. and Miyake, H. 1982. A study of the ecology of the house dust mite in dwelling houses. *Tokai J. Exp. Clin Med.* 7: 233-244.

Valdecasas. 2001: A new subfamily of Rhynchohydracaridae (Acari, Hydrachnellae) from the island of Coiba (Panama) with descriptions of new taxa. *J. Nat. History* 35: 1565-1574.

Van Bronswijk, J.E.M.H. 1972. Food preference of pyroglyphid house dust mites (Acari). *Neth. J. Zool.* 22: 335-340.

Van Bronswijk, J.E.M.H. 1973. *Dermatophagoides pteronyssinus* (Trouessart, 1897) in mattress and floor dust in a temperate climate (Acari Pyroglypoidae). *J. Med. Entomol.* 10: 63-70.

Van Bronswijk, J.E.M.H. 1978. Hausstaubmilben, Workommen und Bedentung. *Allergologie* 1: 55-60.

Van Bronswijk, J.E.M.H 1981. *House dust biology for allergists, Acarologists and mycologists.* NIB Publishers, Zeist, Netherlands: 316pp.

Van de Bund, C.F. 1972. Some observations on predatory action of mites on nematodes. *Zeszyty Problemoux Postepow Nank Roliniczych* 129: 103-110.

Van de Vrie, M. 1973. Studies on prey-predator interactions between *Panonychus ulmi* and *Typhlodromus (A.) potentillae* (Acarina: Tetranychidae, Pbytoseiidae) on apple in the Netherlands. In: *FAO Conference on Ecology in relation to pest control,* Rome, FAO, Rome: 145-160.

Van der Geest, L.P.S., Bosse, Th.C. and Veerman, A. 1983. Development of meridic diet for the two-spotted spider mite, *Tetranychus urticae*. *Entomol. Exp. Appl.* 33: 297-302.

Van Emden, H.F. 1989. *Pest Control.* Cambridge University Press: 116 pp.

Van Hage-Hamsten, M., Machado, L., Barros, M.T. and Johansson, S.O.D. 1990. Comparison of clinical significance and allergenic cross-reactivity of storage mites, *Blomia kulagini* and *Lepidoglyphus destructor Allergy.* 45: 409-417.

Van Lenteren, J.C. and Woets, J. 1988. Biological and integrated pest control in greenhouses. *Ann. Rev. Entomol.* 33: 239-269.

Veerman, A. 1977. Aspects of the induction of diapause in a laboratory strain of the mite, *Tetranychus urticae*. *J. Insect Physiol.* 23: 703-711.

Veerman, A. 1994. Photoperiodic and thermoderodic control of diapause in plant inhabiting mites: A review. *Netherlands J. Zool.* 44(1-2): 139-155.

Vereau, T.W.V., Cueva C.M. and Ojeda P.D. 1978. Biology of cotton red spider mite *Tetranychus cinnabarinus* (Boisduval) (Acarina: Tetranychidae). *Revista PeryabaOdeO Entomologia* 21: 50-54.

Von Torne, E. 1962. Tullgren funnel modification to prevent contamination of the collecting tube with sample debris. In: *Progress in Soil Zoology* (ed. Murphy, P.W.), Butterworth, London: 179-181.

Voorhorst, R., Spieksma, F.T.M. and Varekamp, N. 1969. House Dust Atopy and the House Dust Mite, *Dermatophagoides pteronyssinus* (Trowessart, 1897). *Stafleu's Scientific, Leiden:* 159.

Voorhorst, R., Spieksma-Boezeman, M.I.A. and Spieksma, F.T.M. 1964. Is a mite (*Dermatophagoides* sp.) the producer of the house-dust allergen. *Allergie Asthma* 10: 219-334.

Vorce, C.M. 1881. Wholesome destruction of acari by a fungus. In: *Proc. Am. Soc. Microsc.* 4: 49-50.

Waite, G.K. and McAlpine, J. 1992. Honeybees as carriers of Iychee erinose mite *Eriophyes litchii* (Acari: Eriophyiidae). *Exp. App. Acarol.* 15: 299-302.

Waite, G.K. 1999. New evidance further incriminates honeybees as vectors of lychee erinose mite *Aceria litchii* (Acari: Eriophyiidae). *Exp. App. Acarol.* 23: 145-147.

Walde, S.J. 1991. Patch dynamics of a phytophagous mite population: effect of number of subpopulations. *Ecol.* 72: 1591-1598.

Walde, S.J., Magagula, C.N. and Morton, M.L. 1995. Feeding preference of *Zetzellia mali*: does absolute or relative abundance of prey matter more?

Walia, K.K. 1992. *Studies on nematophagous mites inhabiting agricultural soils and their Bioecological interactions with nematodes.* Ph.D Thesis, CCSHAU, Hisar: 124 pp.

Walia, K.K. 2003. Bioecological interactions between nematophagous mites and plant parasitic nematodes In: *Mites, their identification and management* (eds. Yadav, P.R, Chauhan, R, Putatunda, B.N. and Chillar, B.S.) Department of Entomology, CCS Haryana Agricultural University, Hisar: 190-194.

Walia, KK and Mathur, S. 1994. Predatory potential of two nematophagous mites on fungivorous nematode, *Aphelenchoides composticola in vitro. Indian J. Nematol.* 24: 243-245.

Walia, K.K. and Mathur, S. 1995. Predatory behavior of two nematophagous mites, *Tyrophagus putrescentiae* and *Hypoaspis calcuttaensis*, on root knot nematode, *Meloidogyne javanica*. *Nematol. Medit.* 23: 255-261.

Walia, KK and Mathur, S. 1996. Potential of two nematophagous mites in suppressing root–knot nematode, *Meloidogyne javanica* on Okra. *Indian. J. Nematol.* 126: 172-175.

Walia, K.K. and Mathur, S. 1997. Influence of edaphic factors on soil accarines and plant parasitic nematodes and their interaction. *Indian J. Nematol.* 27(1): 46-52.

Wallace, M.M.H. 1954. The effect of DDT and BHC on the population of the lucerne flea, *Sminthurus viridis* (L.) (Collembola), and its control by predatory mites, *Biscirus* spp. (Bdellidae). *Aust. J. Agric. Res.* 5: 148-155.

Wallace, M.M.H. 1972. A portable power operated apparatus for the field collecting of epigaeic collembo1a and acari. *J. Aust. Entomol. Soc.* 11: 261-263.

Wallace, M.M.H., Tyndale-Biscoe, M. and Holm, E. 1979. The influence of *Macrocheles glaber* on the breeding of the Australian bushfly, *Musca vetustissima* in cowdung. In: *Recent Advances in Acarology* (ed. Rodriquez, J.G.) 2: 217-222.

Wallace, M.M.H. and Holm, E. 1983. Establishment and dispersal of the introduced predatory mite, *Macrocheles peregrimus* Krantz, in Australia. *J. Aust. Entomol. Soc.* 22: 345-348.

Wallace, M.M.H. and Walters, M.C. 1974. The introduction of *Bdellodes lapidaria* (Acari: Bdellidae) from Australia into South Africa for the biological control of *Sminthurus viridis*. *Aust. J. Zool.* 22: 505-517.

Walter, D.E. 1987. Life history, trophic behaviour and description of *Gamasellodes vermivorax* n. sp. (Mesostigmata: Ascidae), a predator of nematodes and arthropods in semiarid grassland soils. *Can.J. Zool.* 65: 1689-1695.

Walter, D.E. 1988a. Nematophagy by soil arthropods from the short grass steppe, Chihuahuan desert and rocky mountains of the central United States. *Agric.Ecosys. Environ.* 24: 307-316.

Walter, D.E. 1988b. Predation and mycophagy by endeostigmatid mites (Acariformes: Prostigmata). *Exp. Appl. Acarol.* 4: 159-166.

Walter, D.E. and lkonen, E.K. 1989. Species, guilds and functional groups: taxonomy and behaviour in nematophagous orthropods..*J. Nematol.* 21: 315-327.

Walter, D.E. and Kaplan, D.T. 1990. Feeding observations on two astigmated mites, *Schwiebea rocketti* (Acaridae) and *Histiostoma bakeri* (Histiomatidae) associated with citrus feeder roots. *Pedobiologia* 34: 281-286.

Walter, D.E. and Kaplan, D.T. 1991. Observations on *Coleoscirus simplex* (Acarina: Prostigmata) a predatory mite that colonizes greenhouse cultures of root-knot nematode *(Meloidogyne* sp.) and a review of feeding behaviour in Cunaxidae. *Exp. Appl. Acarol.* 12: 47-59.

Walter, D.E. and Lindquist, E.E. 1995. The distributions of parthenogenetic ascid mites (Acari: Parasitiformes) do not support the biotic uncertainty hypothesis. *Exp. App. Acarol.* 19: 423-442.

Walter, D.E. and Proctor, H.C. 1999. *Mites: Ecology, Evolution and Behaviour*. University of New South Wales Press, Sydney

Walter, D.E., Hudgens, R.A. and Freckman, D.W. 1986. Consumption of nematodes by fungivorous mites, *Tyrophagus* spp. (Acarina: Astigmata: Acaridae). *Oecologia* 70: 357-361.

Walter, D.E., Hunt, H.W. and Elliott, E.T. 1987. The influence of prey type on the development and reproduction of some predatory soil mites. *Pedobiologia* 30: 419-424.

Walter, D.E., Kaplan, D.T. and Davis, E.L. 1993. Colonization of greenhouse nematode cultures by nematophagous mites and fungi. *Supplement to J. Nematol.* 25(45): 789-794.

Weires, R.W. and Smith, G.L. 1978. Apple mite control, Hydson valley, 1977. *Insectic. Acaria. Tests* 3: 42-43.

Weiser, J. 1963. Uber Massenzuchten von Pyemotes-milben. *Beil Z. entomol.* 13: 547-551.

Welbourn, W.C. 1983. Potential use of trombidioid and erythraeoid mites as biological control agents of insect pests. In: *Biological control of pests by mites* (eds. Hoy, M.A., Cunningham, G.L., Kuntson, L.), Univ. of California: 103-140.

Welty, C., Reissig, W.H., Dennehy, T.J. and Weires, R.W. 1987. Cyhexatin resistance in New York populations of European red mite (Acari: Tetranychidae). *J. Econ. Entomol.* 80: 230-236.

Wermelinger, B., Oertli, J.J. and Delucchi, V. 1985. Effect of host plant nitrogen fertilization on the biology of the two spotted spider mite, *Tetranychus urticae*. *Entomol. Exp. App.* 38: 23-28.

Wharton, G.W. 1976. House dust mites. *J. Med. Entomol.* 12: 577-621.

Wheeler, T.A. and Threlfall, W. 1986. Observations on the ectoparasites of some new found land passerines (Aves: Passerfames). *Can. J. Zool.* 64: 630-636.

White, N.D.G. and Laing, J.E. 1977. Some aspects of the biology and a laboratory life table of the acarine predator, *Zetzellia mali*. *Can Entomol.* 109: 1275-1281.

Willis, R.R. and Axtell, R.C. 1968. Mite predators of the housefly: a comparison of *Fuscuropoda vegetans* and *Macrocheles muscaedomesticae*. *J. Econ. Entomol.* 61: 1669-1674.

Wilson, LT. 1992. *Pest status and ecology of two spotted spider mites on cotton in Austalia and implications for management*. Ph.D Thesis, Univ. Queensland, Brisbane, Australia: 261 pp.

Winston, P.W. and Bates, D.H. 1960. Saturates solutions for the control of humidity in biological research. *Ecology* 41: 232-237.

Wise, G.U., Hennessey, M.K. and Axtell RC. 1988. A new species of manure inhabiting mite in the genus *Poecilochirus* (Acari: Mesostigmata: Parasitidae) predacious on house fly eggs and larvae. *Ann. Entomol. Soc. Am* 81: 209-224.

Witalinski, W., Szlendak, E. and Boczek, J. 1990. Anatomy and ultrastructure of the reproductive systems of *Acarus siro* (Acari: Acaridae). *Exp. App. Acarol.* 10: 1-32.

Wolfenbarger, D.A.1965. Tomato, *Lycopersicon esculentum* and *Lycopersicon* species and genetic markers in relation to mite. *J. Econ. Entomol.* 58: 891-893.

Womersley, H. 1933. A possible biological control of the clover spring tail or Lucerne flea (*Sminthurus viridis* L.) in Western Australia. *J. Aust. Counc. Sci. Indust. Res.* 6: 83-91.

Woodcock, A.A. and Cunnington, A.M. 1980. The allergic importance of house dust and storage mites in asthmatics in Brunei SE. *Asia Clin. Allergy*. 10: 609-615.

Woodring, J.P. 1969. Observations on biology of six species of acarid mites. *Ann. Ent. Soc. Am.* 62: 102-108.

Woodroffe, G.E. 1953. An ecological study of the insects end mites in the nests of certain birds in Britain. *Bull. Ebntomol. Res.* 44: 739-772.

Woolley TA. 1988. Acarology–Mites and Human Welfare. John Wiley and Sons, New York. 484 pp.

Woyke, J. 1985. *Tropilaelaps clareae,* a serious pest of *Apis mellifera* in the tropics, but not dangerous for apiculture in temperate zones. *Am. Bee J.* 125:497-99.

Woyke, J. 1987 a. Infestation of honey bee (Apis *mellifera*) colonies by the parsitic mites, *Varroa jacobsoni* and *Tropilaelaps clareae* in South Vietnam and results of chemical treatment. *J. apic. Res.* 26(1): 64-67.

Wraith, D.G., Cunnington, A.M. and Seymour, W.M. 1979. The role and allergenic importance of storage mites in house dust and other environments. *Clin. Allergy.* 9: 545-561.

Yadav, G.S., Anand, R.K. and Yadav, P.R. 1987. Bioefficacy of some chemicals against red spider mite on brinjal. *Indian J. Ent.* 49(4): 582-584.

Yamada, T., Yoneda, H. and Mitsuo, A. 1986. Resistance to benzoate in mites. In: *Pest resistance to pesticides* (eds. Georghio, G.P. and Saito, T.), Plenum Publ., New York: 445-451.

Yaninek, J.S. and Herren, H.R. 1988. Introduction and spread of the cassava green mite, *Mononychellus tanajoa* (Bander) (Acari: Tetranychidae) an exotic pest in Africa and the search for appropriate control methods: a review. *Bull. Ent. Res.* 78: 1-13.

Yaninek, J.S., Baumgartner, J. and Gutierrez, A.P. 1991. Sampling *Mononychellus tanajoa* (Acari: Tetranychidae) on cassava in Africa. *Bull Ent. Res.* 81: 201-208.

Yoshizawa, T., Yamamoto, T. and Yamamoto, R. 1971. Synergestic attractancy of cheese components for chese mite *Tyrophagus putrescentiae* (Schrank). *Batyn-Kagaku* 36(1): 1-6.

Young, R.P., Hart, B.J., Faux, J.A. and Hopkin, J.M. 1990. House dust mite sensitivity: a comparison of immune response with exposure to four species of pyroglyphidae. *Clin. Exp. Allergy.* 20: 319-325.

Yousef, A.A., El-Enany, M.A. and El-Rahman, S.LA. 1992. Effect of prey species on the biology of the Cheyletid mite *Acaropsis docta* Berlese. *Egyptian J. Agri. Res.* 70 (2): 449-506.

Yu, H. and Shi, J.M. 1991. Studies on *Shizotetranychus nanjingensis*. *J. Bamboo Res.* 10(2): 61-67.

Yunker, C.E. 1973. Parasites of endothermal laboratory animals. Chapter 15, Mites. In: *Parasites of Laboratory Animals* (ed. Flynn, R.J.), Iowa State Univ. Press. Ames: 425-492.

Zacharda, M., Pultar, O. and Muka, J. 1988. Washing technique for monitoring mites in apple orchards. *Exp. Appl. Acarol.* 5: 181-183.

Zaher, M.A., Soliman, Z.R. and El-Bishlawy, S.M. 1975. Feeding habits of the predaceous mite, *Cunaxa capreolus* (Acarina: Cunaxidae). *Entomophaga* 20: 209-212.

Zaher, M.A., Soliman, Z.R. and Rakha, M.A. 1984. Biological studies on the predatory mite *Saniosulus nudus* Summers (Raphignathoida: Eupalopsellidae). In: *Acarology* VI (eds. Griffiths, D.A. and Bowman, C.E.), Ellis Horwood, Chichester 1: 597-600.

Zdarkova, E. 1986. Mass rearing of the predator *Cheyletus eruditus* (Schrank) (Acarina: Cheyletidae) for biological control of acerid mites infesting stored products. *Crop Prot.* 5: 122-124.

Zdarkova, E. 1996. *Modern Acarology*, SPB Publ., Czec Repubolic: 607-610.

Zdarkova, E. and Horak, E. 1987. Contact acaricides may not restrain effectiveness of the biological control against stored food mites. *Acta Entomol. Bohemolov* 84: 414-421.

Zdarkova, E. and Pulpan, J. 1973. Low temperature storage of the predatory mite, *Cheyletus eruditus* (Schrank) for future use in biological control. *J. Stored Prod. Res.* 9: 217-220.

Zdarkova, E. and Stehlikove, B. 1981. The development of stored food mites on green coffee, Cocoa beans had legumes. *SR Urtz (Ustave Vedeckotech Inf Zemed) Octiv Rostl.* 17(3): 217-224.

Zeman, P. and Zelegny, J. 1985. The susceptibility of the poultry red nite, *Dermanyssus gallinae* (De Geer, 1978) to some acaricides under laboratory conditions. *Exp. Appl. Acarol.* 1: 17-22.

Zhang, Y.X., Liu, Q.Y., Lin, J.Z., Song, M.G., Cai, Q.J. and He, X.Y. 1998. Studies on species distribution and damage by spider mites in bamboo trees in Fujian Province. *J. Fujiian Acad. Agric. Sci.* 12(3): 11-15.

Zhang, Z.Q. 1988. Progresses and future prospects of trombidioidca and Erythraeoidea mites as candidates for biological control of insect pests. *Chin..J. Biol. Ctrl.* 4: 79-82.

Zhang, Z.Q. 1991. Parasitism of *Acyrthosiphon pisum* by *Allothrombium pulvinum* (Acariformes: Trombidiiae): Host attachment site, host size selection, super, parasitism and effect on host. *Exp. Appl. Acarol.* 11: 137-147.

Zhang, Z.Q. 1991. Biology of mites of Allothrumbiinae (Acari:Trombidiida) and their potential role in pest control. In: *Modern Acarology* (eds. Dusbabek, F. and Bukva, V.), Academia, Prague and SPB Academic Publishing BV, The Hogue. 2: 513-520.

Zhang, Z.Q. and Croft, B.A. 1995. Intraspecific competition in immature *Amblyseius fallacis, Amblyseius andersoni, Typhlodromus occidentalis and Typhlodromus pyri* (Acari: Phytoseiidae) *Exp. Appl. Acarol*. 19: 65-77.

Zhang, Z.Q. and Xin, J.L. 1989. Biology of *Allothrombium pulvinnum* (Acariformes: Trombidiidae), a potential biological control agent of aphids in China. *Exp. Appl. Acarol.* 6: 101-108.

Zhang, Z.Q. and Xin, J.L. 1989. Studies on the morphology and life history of *Allothrombium pulvinum* Ewing (Acariformes: Trombididae) *Acta Entomol. Sinica.* 32(2): 192-199.

Zhang, Z.Q. and Saboori, A. 1996. A new host record of *Allothrombium ovatum* larvae (Acari: Trombidiidae) ectoparasitic on *Coccinella septempunctata* (Coleoptera: Coccinellidae). *Sys. Appl. Acarol.* 1: 207-208.

Zhang, Z.Q. Chen, P.R. Wang, K. and Wang, X.Y. 1993. Overdispersion of *Allothrombium pulvinum* larvae (Acari: Trombidiidae) parastic on *Aphis gossypii* (Homoptera: Aphididae) in cotton fields. *Ecol. Entomol.* 18: 379-384.

Zhao, G., Liu, W. and Knowles, C.O. 1996. Clofentezine toxicity and fate in the bulb mite *Rhizoglyphus echinopus* (Acari: Acaridae). *Exp. App. Acarol.* 20: 215-222.

Zhou, G.A., Zou, J.J. and Huang, S.D. 1994. The effect of *Agaratum conyzoides* interplanting in hilly citrus orchards in Human province on citrus mite and insect populations. *Acta Phytophylacica Sinica* 21: 57-61.

Zumpt, F. 1961. The arthopod parasites of vertebrates in Africa south of Sahara Ethiopian region) *S. Afr. Inst. Med. Res. Johannesburg* 1: 1-457.

Index